I0044605

Signals and Systems:
An Engineering Perspective

Signals and Systems: An Engineering Perspective

Andrew Burton

Signals and Systems: An Engineering Perspective
Andrew Burton
ISBN: 978-1-64172-398-5 (Hardback)

Published by Larsen and Keller Education,
5 Penn Plaza,
19th Floor,
New York, NY 10001, USA

Cataloging-in-Publication Data

Signals and systems : an engineering perspective / Andrew Burton.
p. cm.
Includes bibliographical references and index.
ISBN 978-1-64172-398-5
1. Signal processing. 2. System analysis. 3. Signal theory (Telecommunication).
4. Engineering. 5. Systems engineering. I. Burton, Andrew.
TK5102.9 .S54 2020
621.382 2--dc23

For more information regarding Larsen and Keller Education and its products, please visit the publisher's website www.larsen-keller.com

Table of Contents

Preface

This book has been written, keeping in view that students want more practical information. Thus, my aim has been to make it as comprehensive as possible for the readers. I would like to extend my thanks to my family and co-workers for their knowledge, support and encouragement all along.

A signal is a function that gives information about a phenomenon. The field of electrical engineering that studies output and input signals, and mathematical representations between systems is known as signals and systems. The four main domains of signals and systems are frequency, time, s and z. It is a subset of mathematical modeling. Signal processing involves analyzing, synthesizing and modifying signals. Its techniques are used to improve efficiency and subjective quality, and transmission. It receives signals as well as produces them. System is a physical set of components. It has one or more input and output signals. In signals and systems, signals are classified according to many criteria. Different types of signals include analog, digital, deterministic, random, energy, power, etc. The book aims to shed light on some of the unexplored aspects of signals and systems. Such selected concepts that redefine the subject have been presented in it. For all those who are interested in signals and systems, this book can prove to be an essential guide.

A brief description of the chapters is provided below for further understanding:

Chapter – Introduction

A signal is defined as an electronic pulse or wave which is transmitted and received. A few of its aspects are analog to digital conversion, discrete-time signal, signal processing, time shifting, time reversal, time scaling, etc. This is an introductory chapter which will briefly introduce all these aspects related to signals.

Chapter – Signal Processing and Quantization

Quantization refers to the process of mapping infinite input values from a large set to a finite set of output values. Digital signal processing, analog signal processing, Nyquist-Shannon sampling theorem, aliasing, etc. fall under the domain of signal processing. This chapter has been carefully written to provide an easy understanding of signal processing and quantization.

Chapter – Systems used in Signal Processing

A field of electric engineering which aims at analyzing, synthesizing and modifying electromagnetic signals such as of sound, images, videos, etc. is called signal processing. It includes significant systems such as lumped parameter and distributed parameter systems, casual and non-casual systems, linear and non-linear systems and discrete time system. This chapter closely examines these systems used in signal processing to provide an extensive understanding of the subject.

Chapter – Fourier Series and Fourier Transform

Fourier series represents an expansion of periodic operation in terms of an infinite sum sines and cosines. Fourier transform converts a general and non-periodic operation into its constituent frequencies. The topics elaborated in this chapter will help in gaining a better perspective about the fourier series and fourier transform.

Chapter - Laplace Transform

A transformation which is used to convert an operation of a real variable (t) into complex variable (s) is termed as laplace transform. It is used for analysing and developing circuits such as filters. This chapter delves into various concepts such as laplace transform properties, region of convergence, the laplace transform of a function, existence of laplace transform, etc. which will provide in-depth knowledge of the subject.

Chapter - Z-Transform

In signal processing, Z-transformation is used to transform a series of real or complex values into a complex frequency domain representation. Region of convergence, properties of Z transform, pole zero plot, inverse Z transform, etc. are some of the principles associated with it. This chapter discusses these principles related to Z Transform in detail.

Andrew Burton

1

Introduction

A signal is defined as an electronic pulse or wave which is transmitted and received. A few of its aspects are analog to digital conversion, discrete-time signal, signal processing, time shifting, time reversal, time scaling, etc. This is an introductory chapter which will briefly introduce all these aspects related to signals.

Signals

A signal is defined as any physical or virtual quantity that varies with time or space or any other independent variable or variables.

Graphically, the independent variable is represented by horizontal axis or x-axis. And the dependent variable is represented by vertical axis or y-axis.

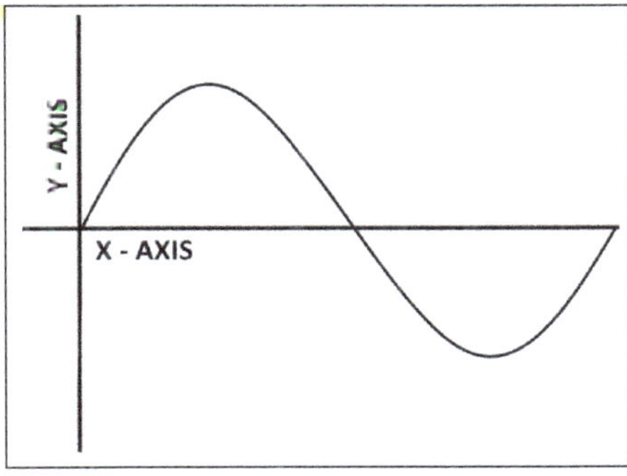

Mathematically, a signal is a function of one or more than one independent variables.

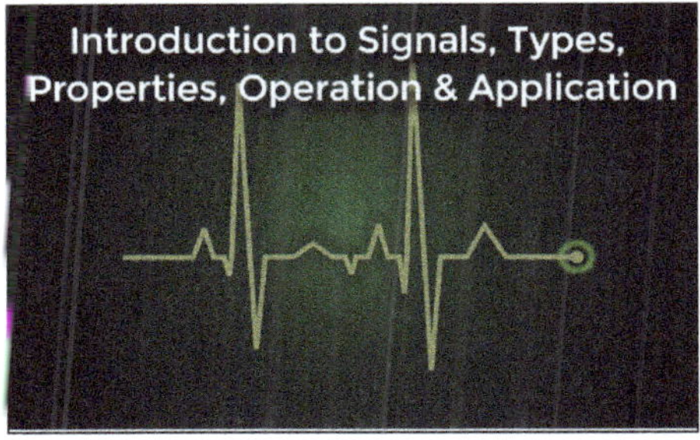

Single Variable Signal

It depends on a single independent variable. It either varies linearly or non-linearly depending on the expression of the signal. Examples of single variable signal are:

$$S(x) = x + 5$$

$S(x) = x^2 + 5$ Where x is the variable,

$S(t) = \cos(wt + \theta)$ Where t is the variable.

Two Variables Signal

A two-variable signal varies with the change in the two independent variables. Example of a two-variable signal is:

$$S(x,y) = 2x + 5y$$

Characteristics of Signal

A signal is defined by its characteristics. It shows the nature of the signal. These characteristics are given below:

Amplitude

Amplitude is the strength or height of the signal waveform. Visually, it is the height of the waveform from its centerline or x-axis. The y-axis of a signal's waveform shows the amplitude of a signal. The amplitude of a signal varies with time.

For example, the amplitude of a sine wave is the maximum height of the waveform on Y-axis.

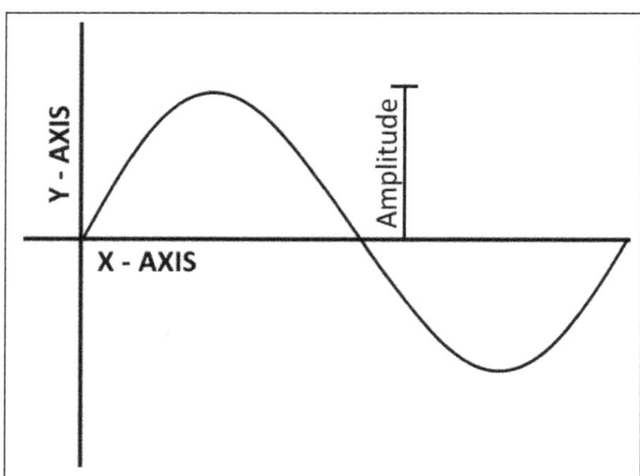

The signal's strength is usually measured in decibels db.

Frequency

Frequency is the rate of repetitions of a signal's waveform in a second.

Periodic signals repeat its cycle after some time. The number of cycles in a second is known as Frequency. The unit of Frequency is hertz (Hz) and one hertz is equal to one cycle per second. It is measured along the x-axis of the waveform.

For example, a sine wave of 5 hertz will complete its 5 cycles in a one second.

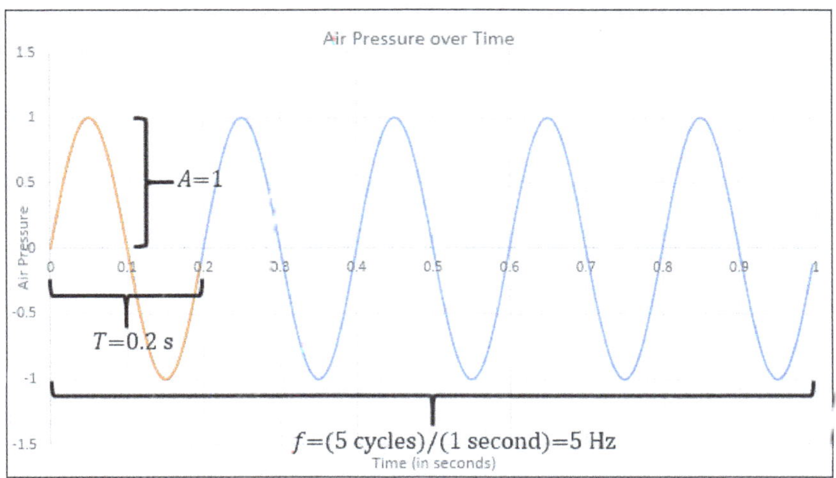

Time Period

The time period of a signal is the time in which it completes its one full cycle. The unit of the time period is Second. The time period is denoted by 'T' and it is the inverse of frequency. i.e.

T=1/F

For example, a sine wave of time period 10 sec will complete its one full cycle in 10 seconds.

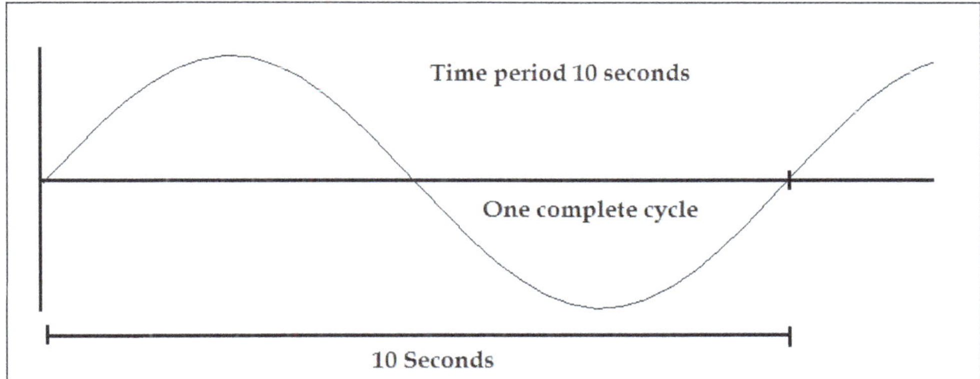

Phase

The phase of a sinusoidal signal is the shift or offset in its origin or starting point. The phase shift can be lagging or leading. Usually, the original sinusoidal signals have 0° degree phase and start at 0 amplitude but an offset in phase will shift its starting amplitude to other than 0.

An example of 45° phase shift is given below. The signal remains the same but its origin is shifted to 45°.

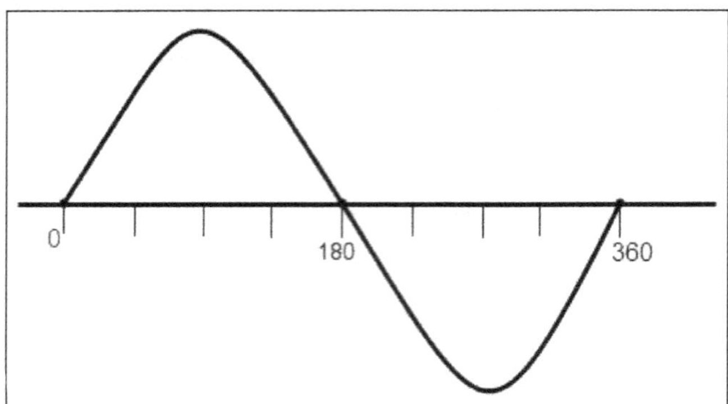

The phase shift can be from 0° to 360° in degrees or 0 to 2π in radians. 360° degree or 2π radians is one complete period.

Signal Size

The size of a signal is a number that shows the strength or largeness of that signal. As we know, a signal's amplitude varies with respect to time. Because of this variation, we cannot say that its amplitude can be its size. To measure the signal size, we have to take into account the area covered by the amplitude of the signal within the time duration.

According to the size of the signal, there are two parameters.

Signal Energy

The energy of the signal is the area of the signal under its curve. But the signal can be in both positive and negative region. Due to which, it will cancel each other's effect resulting in a smaller signal. To eradicate this problem, we take the square of the signal's amplitude which is always positive.

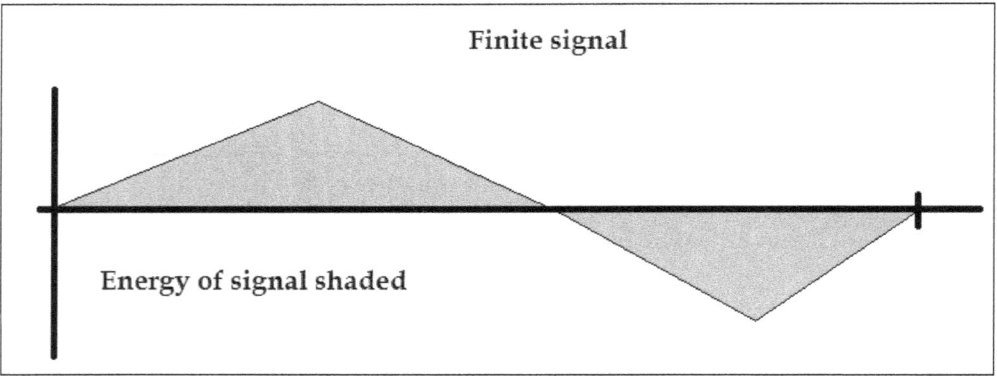

For a signal g(t), the area under the $g^2(t)$ is known as the Energy of the signal.

$$E_g = \int_{-\infty}^{\infty} g^2(t)dt$$

Unit of Energy of Signal

This energy is not taken as in its conventional sense, but it shows the signal size. Therefore, its unit is not joule. The unit of energy depends on the signal. If it is a voltage signal then its unit will be volts2/second.

Limitation

The energy of a signal can be measured only if the signal is finite. The infinite signal will have infinite energy, which is absurd. A finite signal's amplitude goes to 0 as the time (t) approaches to infinity (∞).

So it is necessary that the signal is a finite signal if you want to measure its energy.

Signal Power

If the signal is an infinite signal i.e. its amplitude does not go to 0 as time t approaches to ∞, we cannot measure its energy. In such a case, we take the time average (Time period) of the energy of the signal as the power of the signal.

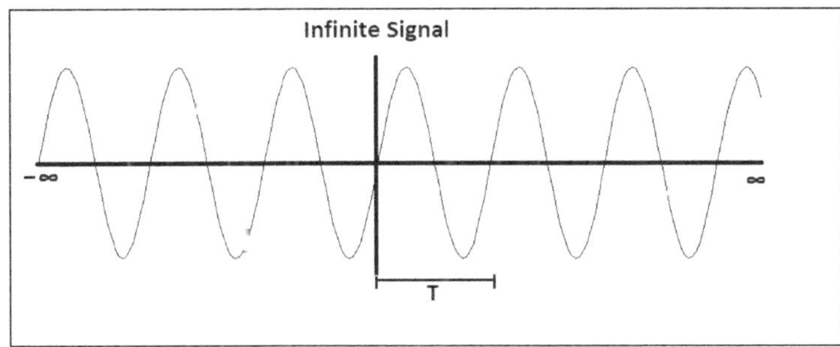

$$P_g = \lim_{T\to\infty} \frac{1}{T} \int_{-T/2}^{T/2} g^2(t)dt$$

Unit of Power

Similar to Energy of the signal, this power is also not taken in the conventional sense. It will also depend on the signal to be measured. If the signal is a voltage signal, then the power will be in volts^2.

Limitation

Just like the energy of the signal, the measurement of the power of a signal also has some limitation that the signal must be of a periodic nature. An infinite and non-periodic signal neither have energy nor power.

Operation of Signal

Some basic operation of signals are given below.

Time Shifting

Time-shifting means movement of the signal across the time axis (horizontal axis). A time shift in a signal does not change the signal itself but only shifts the origin of the signal from its original point along time-axis.

Basically, addition in time is time shifting. To time-shift a signal g(t), t should be replaced with (t-T), where T is the seconds of time-shift. Therefore, g(t-T) is the time-shifted signal by T seconds.

Time shift can be right-shift (delay) or left-shift (advance).

If the time-shift T is positive than the signal will shift to the right (delay). For example, the signal g(t-4) is the shifted version of g(t) with 4 seconds delay.

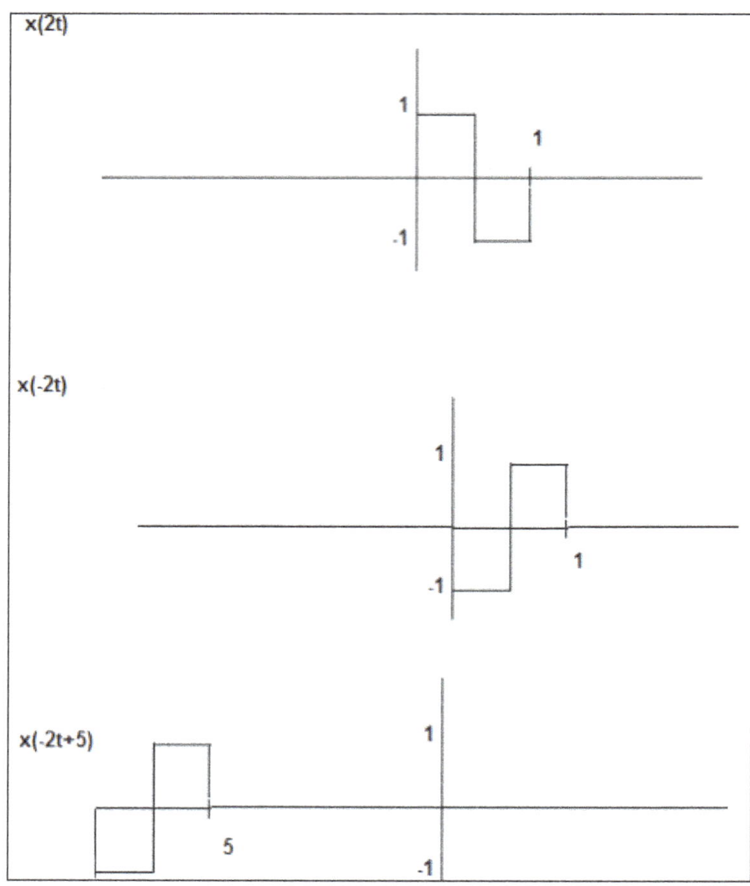

If the time-shift T is negative than the signal will shift to the left (advance). The signal g(t+4) is the shifted version of g(t) with 4 seconds to the left.

Time Scaling

Time scaling of a signal means to compress or expand the signal. It is achieved by multiplying the time variable of the signal by a factor. The signal expands or compresses depending on the factor. Suppose a signal g(t) than its scaled version is g(at).

If the factor a>1 then the signal will compress. And the operation is called signal compression. Compressing a signal will make the signal fast as it becomes smaller and its time duration become less.

If a<1 then the signal will expand. And the operation is called signal dilation.

After scaling, the origin of the signal remains unchanged. Expanding the signal will make the signal slow as it becomes wider and covers more time duration.

Time Inversion

In time inversion, the signal is flipped about the y-axis (vertical axis). The resultant signal is the mirror image of the original signal.

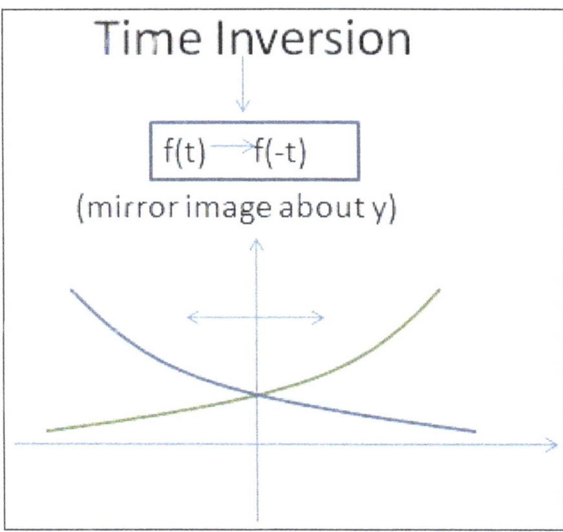

Time inversion is a special case of time-scaling in which the factor a=-1. Therefore to invert a signal, we replace it's (t) with (-t).

Mathematically, the time-invert of signal g(t) is g(-t).

Analog Signal

Analog is best explained by the transmission of signal such as sound or human speech, over an electrified copper wire. In its native form, human speech is an oscillatory disturbance in the air. Which varies in terms of its volume or power (amplitude) and its pitch or tone (frequency)? Analogous variations in electrical or radio waves are created in order to transmit the analog information signal for video or audio or both over a network from a transmitter to a receiver (TV set, computer connected with antenna). At the receiving end an approximation (analog) of the original information is presented.

Information which is analog in its native form can vary continuously in terms of intensity (volume or brightness) and frequency (tone or color). Those variations in the native information stream are translated in an analog electrical network into variations in -the amplitude and frequency of the carrier signal. In other words, the carrier signal is modulated (varied) in order to create an analog of the original information stream.

The electromagnetic sinusoidal (waveform) or sine wave can be varied in amplitude at a fixed frequency, using Amplitude Modulation (AM). Alternatively, the frequency of the sine wave can be varied at constant amplitude using Frequency Modulation (FM). Additionally, both frequency and amplitude can be modulated simultaneously.

- Analog signal can have infinite number of values and varies continuously with time.
- Analog signal is usually represented by sine wave.
- As shown in figure each cycle consists of a single arc above the time axis followed by a single arc below the time axis.
- Example of analog signal is human voice. When we speak, we use air to transmit an analog signal. Electrical signal from an audio tape, can also be in analog form.

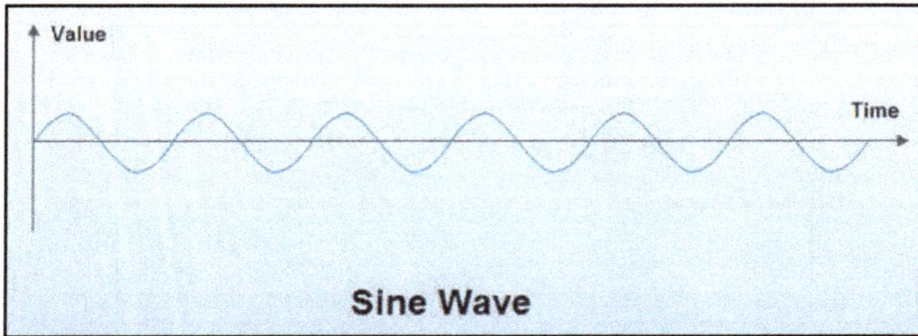

Characteristics of Analog Signal

Amplitude

- Amplitude of a signal refers to the height of the signal.
- It is equal to the vertical distance from a given point on the waveform to the horizontal axis.
- The maximum amplitude of a sine wave is equal to the highest value it reaches on the vertical axis as shown in figure.
- Amplitude is measured in volts, amperes or watts depending on the type of signal. A volt is used for voltage, ampere for current and watts for power.

Period

- Period refers to the amount of time in which a signal completes one cycle.
- It is measured in seconds.
- Other units used to measure period are millisecond (10^{-3} sec.) microsecond (10^{-6} sec), nanosecond (10^{-9} sec) and picoseconds (10^{-12} sec).

Frequency

- It refers to the number of wave patterns completed in a given period of time.
- To be more precise, frequency refers to number of periods in one second or number of cycles per second.

- Frequency is measured in Hertz (Hz)
- Other units used to express frequency are kilohertz (10^3 Hz) Megahertz (10^6 Hz), gigahertz (10^9 Hz) and terahertz (10^{12} Hz).
- Frequency and period are the inverse of each other. Period is the inverse of frequency and frequency is the inverse of period.

Phase

- Phase describes the position of the waveform relative to time zero.
- Phase describes the amount by which the waveform shifts forward or backward along the time axis.
- It indicates the status of first cycle.
- Phase is measured in degrees or radians.
- A phase shift of 3600 indicates a shift of a complete period, a phase shift of 180° indicates a shift of half period and a phase shift of 90° indicates a shift of a quarter of a period as shown in fig. below.

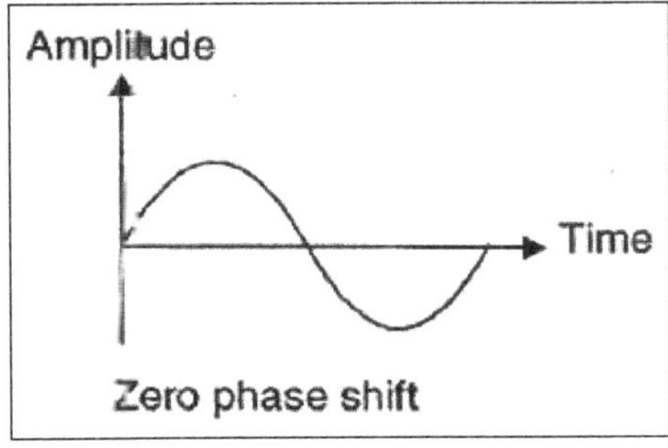

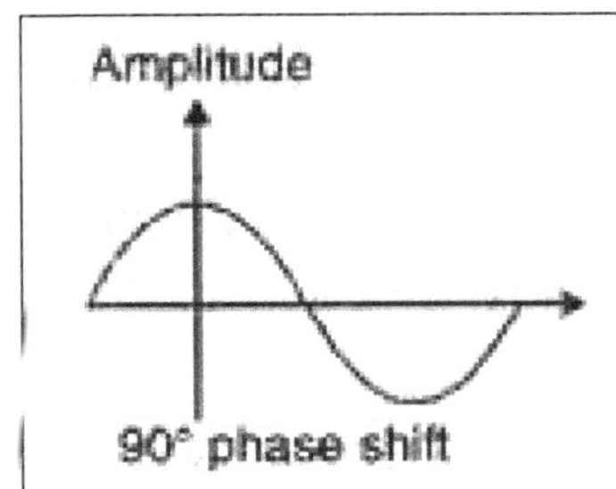

A shift of a quarter of a period or ¼ cycle.

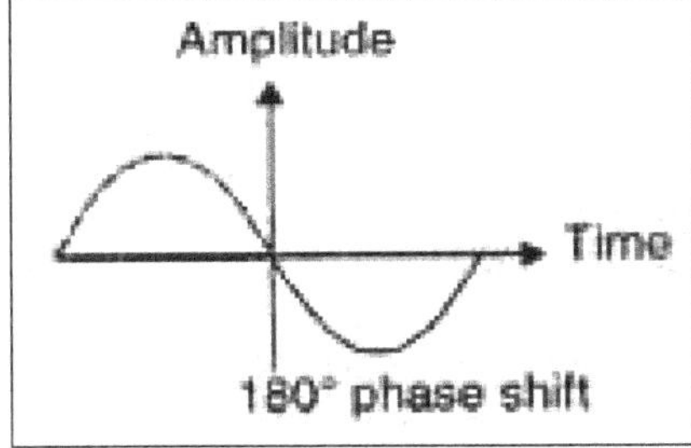

A shift of half period or ½ cycle.

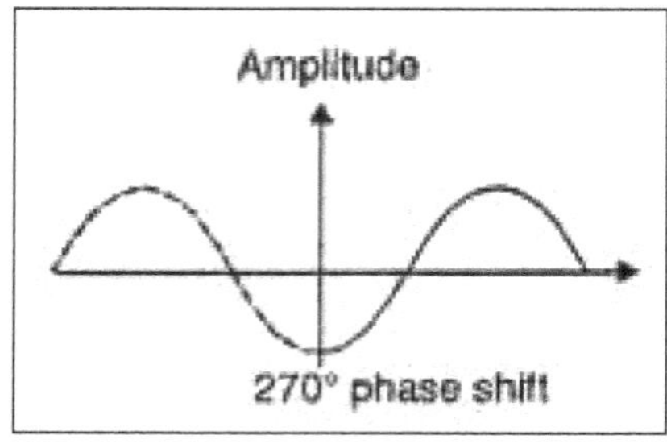

A shift of ¾ cycle.

Advantages of Analog Signals

- Best suited for the transmission of audio and video.
- Consumes less bandwidth than digital signals to carry the same information.
- Analog systems are readily in place around the world.
- Analog signal is less susceptible to noise.

Digital Signal

Computers are digital in nature. Computers process, store, and communicate information in binary form, i.e. in the combination of 1s and 0s which has specific meaning in computer language. A binary digit (bit) is an individual 1 or O. Multiple bit streams are used in a computer network.

Contemporary computer systems communicate in binary mode through variations in electrical voltage. Digital signaling, in an electrical network, 'involves a signal which varies in voltage to represent one of two discrete and well-defined states as depicted in figure such as either a positive (+) voltage and a null or zero (0) voltage (unipolar) or a positive (+) or a negative (-) voltage (bipolar).

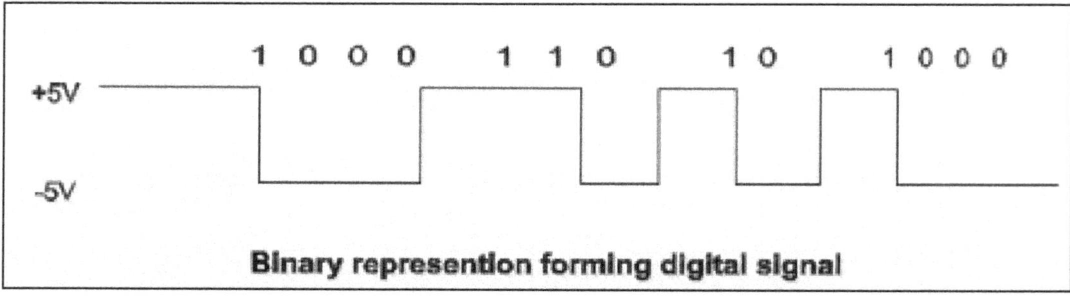

Binary represention forming digital signal

Although analog voice and video can be converted into digital, and digital data can be converted to analog, each format has its own advantages.

- It can have only a limited number of defined values such as 1 and O.
- The transition of a digital signal from one value to other value is instantaneous.
- Digital signals are represented by square wave.

- In digital signals 1 is represented by having a positive voltage and 0 is represented by having no voltage or zero voltage as shown in figure.
- All the signals generated by computers and other digital devices are digital in nature.

Characteristics of Digital Signals

Bit Interval

It is the time required to send one single bit.

Bit Rate

- It refers to the number of bit intervals in one second.
- Therefore bit rate is the number of bits sent in one second.
- Bit rate is expressed in bits per second (bps).
- Other units used to express bit rate are Kbps, Mbps and Gbps.
  - 1 kilobit per second (Kbps) = 1,000 bits per second.
  - 1 Megabit per second (Mbps) = 1,000,000 bits per second.
  - 1 Gigabit per second (Gbps) = 1,000,000,000 bits per second.

Advantages of Digital Signals

- Digital Data: Digital transmission certainly has the advantage where binary computer data is being transmitted. The equipment required to convert digital data to analog format and transmitting the digital bit streams over an analog network can be expensive, susceptible to failure, and can create errors in the information.
- Compression: Digital data can be compressed relatively easily, thereby increasing the efficiency of transmission. As a result, substantial volumes of voice, data, video and image information can be transmitted using relatively little raw bandwidth.
- Security: Digital systems offer better security. While analog systems offer some measure of security through the scrambling of several frequencies. Scrambling is fairly simple to defeat. Digital information, on the other hand, can be encrypted to create the appearance of a single, pseudorandom bit stream. Thereby, the true meaning of individual bits, sets of bits, or the total bit stream cannot be determined without having the key to unlock the encryption algorithm employed.
- Quality: Digital transmission offers improved error performance (quality) as compared to analog. This is due to the devices that boost the signal at periodic intervals in the transmission system in order to overcome the effects of attenuation. Additionally, digital networks deal more effectively with noise, which always is present in transmission networks.

- Cost: The cost of the computer components required in digital conversion and transmission has dropped considerably, while the ruggedness and reliability of those components has increased over the years.
- Upgradeability: Since digital networks are comprised of computer (digital) components, they are relatively easy to upgrade. Such upgrading can increase bandwidth, reduces the incidence of error and enhance functional value. Some upgrading can be effected remotely over a network, eliminating the need to dispatch expensive technicians for that purpose.
- Management: Generally speaking, digital networks can be managed much more easily and effectively due to the fact that such networks consist of computerized components. Such components can sense their own level of performance, isolate and diagnose failures, initiate alarms, respond to queries, and respond to commands to correct any failure. Further, the cost of these components continues to drop.

Analog to Digital Conversion

The following techniques can be used for Analog to Digital Conversion:

Pulse Code Modulation

The most common technique to change an analog signal to digital data is called pulse code modulation (PCM). A PCM encoder has the following three processes:

- Sampling.
- Quantization.
- Encoding.

Low Pass Filter

The low pass filter eliminates the high frequency components present in the input analog signal to ensure that the input signal to sampler is free from the unwanted frequency components.This is done to avoid aliasing of the message signal.

Sampling: The first step in PCM is sampling. Sampling is a process of measuring the amplitude of a continuous-time signal at discrete instants, converting the continuous signal into a discrete signal. There are three sampling methods:

- Ideal Sampling: In ideal Sampling also known as Instantaneous sampling pulses from the analog signal are sampled. This is an ideal sampling method and cannot be easily implemented.
- Natural Sampling: Natural Sampling is a practical method of sampling in which pulse have finite width equal to T.The result is a sequence of samples that retain the shape of the analog signal.

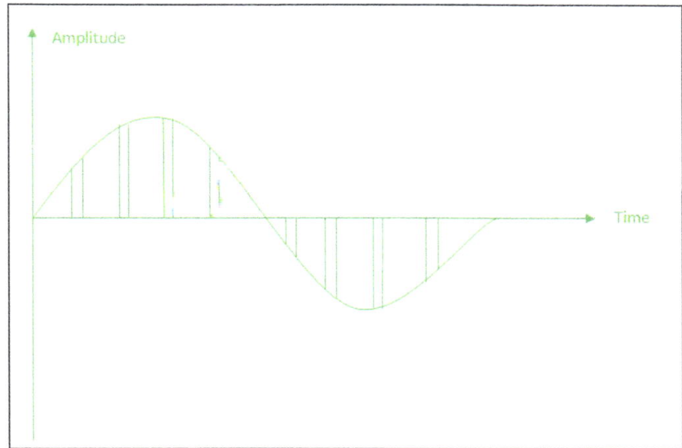

- Flat top sampling: In comparison to natural sampling flat top sampling can be easily obtained. In this sampling technique, the top of the samples remains constant by using a circuit. This is the most common sampling method used.

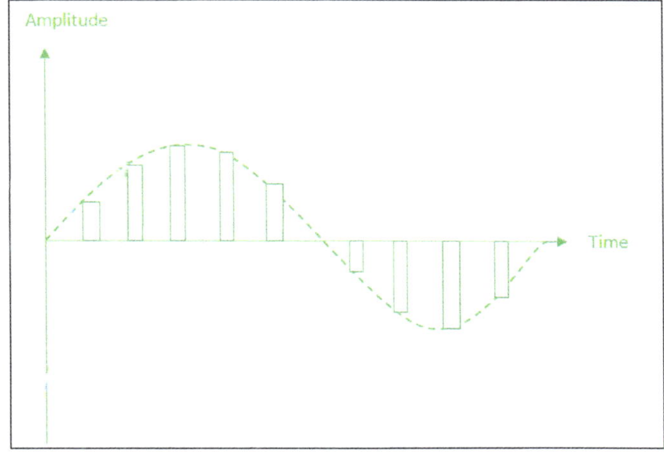

Nyquist Theorem

One important consideration is the sampling rate or frequency. According to the Nyquist theorem, the sampling rate must be at least 2 times the highest frequency contained in the signal. It is also known as the minimum sampling rate and given by:

Fs =2*fh

Quantization

The result of sampling is a series of pulses with amplitude values between the maximum and minimum amplitudes of the signal. The set of amplitudes can be infinite with non-integral values between two limits.

The following are the steps in Quantization:

- We assume that the signal has amplitudes between Vmax and Vmin.

- We divide it into L zones each of height d where:

  d= (Vmax- Vmin)/ L

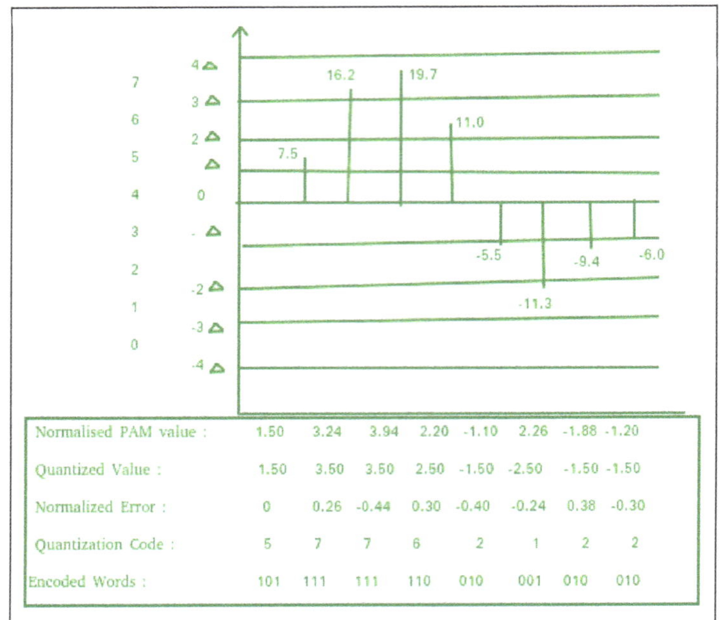

Normalised PAM value :	1.50	3.24	3.94	2.20	-1.10	2.26	-1.88	-1.20
Quantized Value :	1.50	3.50	3.50	2.50	-1.50	-2.50	-1.50	-1.50
Normalized Error :	0	0.26	-0.44	0.30	-0.40	-0.24	0.38	-0.30
Quantization Code :	5	7	7	6	2	1	2	2
Encoded Words :	101	111	111	110	010	001	010	010

- The value at the top of each sample in the graph shows the actual amplitude.
- The normalized pulse amplitude modulation(PAM) value is calculated using the formula amplitude/d.
- After this we calculate the quantized value which the process selects from the middle of each zone.
- The Quantized error is given by the difference between quantised value and normalised PAM value.
- The Quantization code for each sample based on quantization levels at the left of the graph.

Encoding

The digitization of the analog signal is done by the encoder. After each sample is quantized and the number of bits per sample is decided, each sample can be changed to an n bit code. Encoding also minimizes the bandwidth used.

Delta Modulation

Since PCM is a very complex technique, other techniques have been developed to reduce the complexity of PCM. The simplest is delta Modulation. Delta Modulation finds the change from the previous value.

Modulator – The modulator is used at the sender site to create a stream of bits from an analog signal. The process records a small positive change called delta. If the delta is positive, the process records a 1 else the process records a 0. The modulator builds a second signal that resembles a staircase. The input signal is then compared with this gradually made staircase signal.

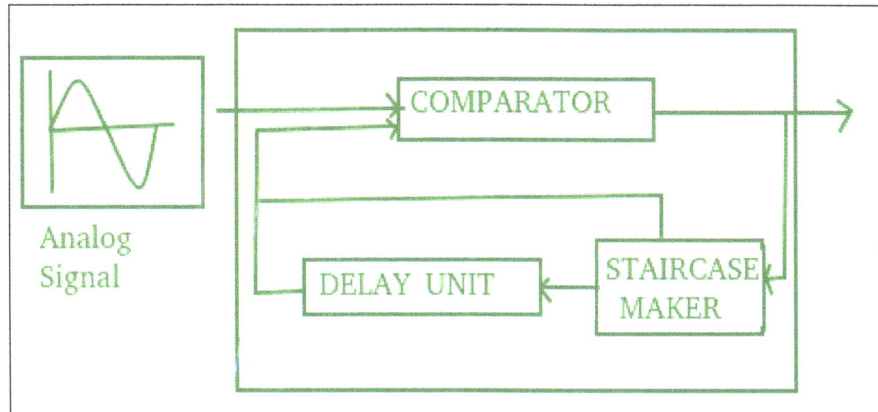

We have the following rules for output:

- If the input analog signal is higher than the last value of the staircase signal, increase delta by 1, and the bit in the digital data is 1.
- If the input analog signal is lower than the last value of the staircase signal, decrease delta by 1, and the bit in the digital data is 0.

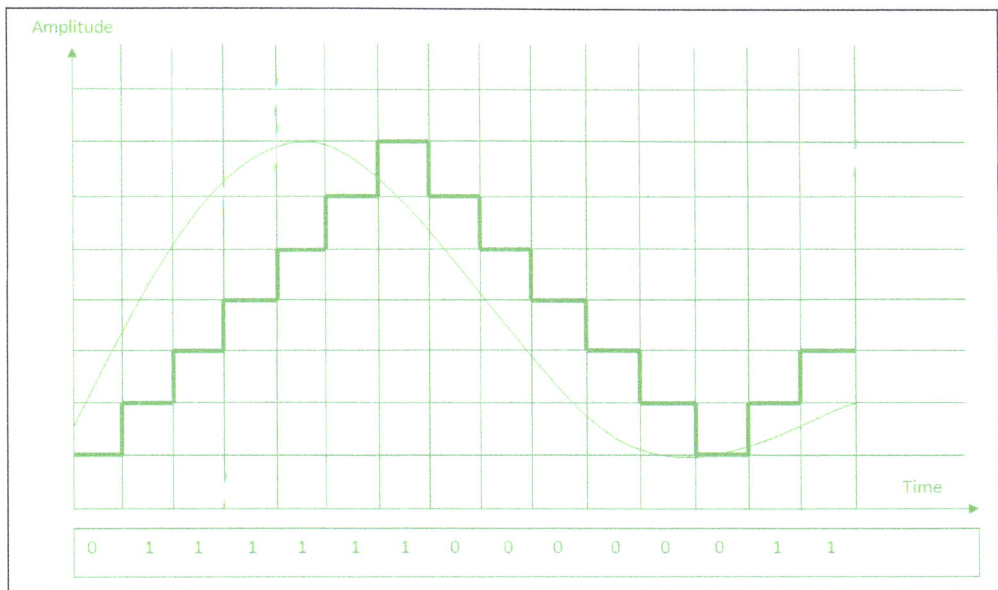

Adaptive Delta Modulation

The performance of a delta modulator can be improved significantly by making the step size of the modulator assume a time-varying form. A larger step-size is needed where the message has a steep slope of modulating signal and a smaller step-size is needed where the message has a small slope. The size is adapted according to the level of the input signal. This method is known as adaptive delta modulation (ADM).

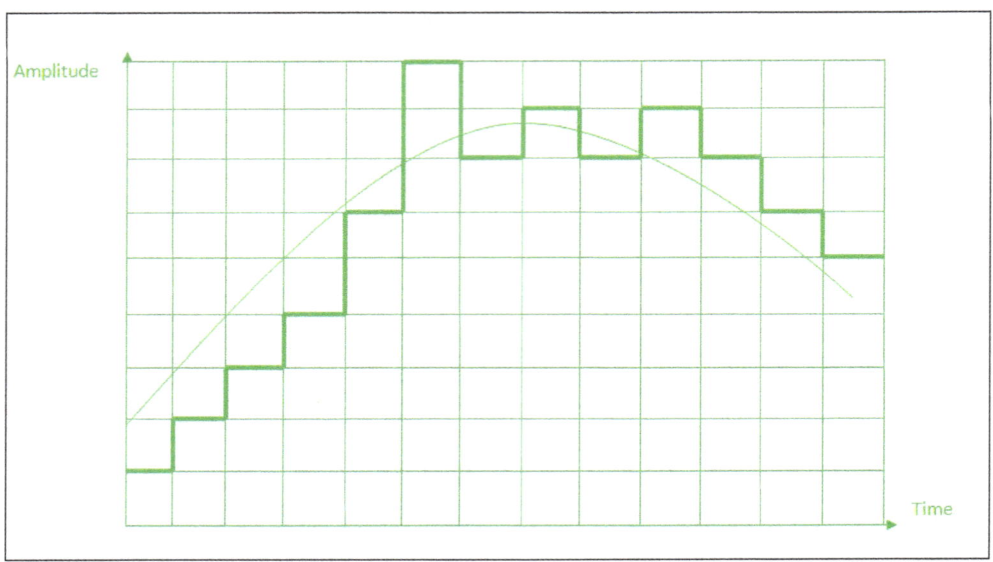

Discrete-time Signal

A discrete-time signal is represented as a sequence of numbers:

$$x = [n] = \{x[n]\}, \qquad -\infty < n < \infty.$$

Here n is an integer, and $x[n]$ is the nth sample in the sequence.

Discrete-time signals are often obtained by sampling continuous-time signals. In this case the nth sample of the sequence is equal to the value of the analogue signal $x_a(t)$ at time $t = nT$:

$$x[n] = x_a(nT), \quad -\infty < n < \infty.$$

The sampling period is then equal to T , and the sampling frequency is $f_s = 1/T$.

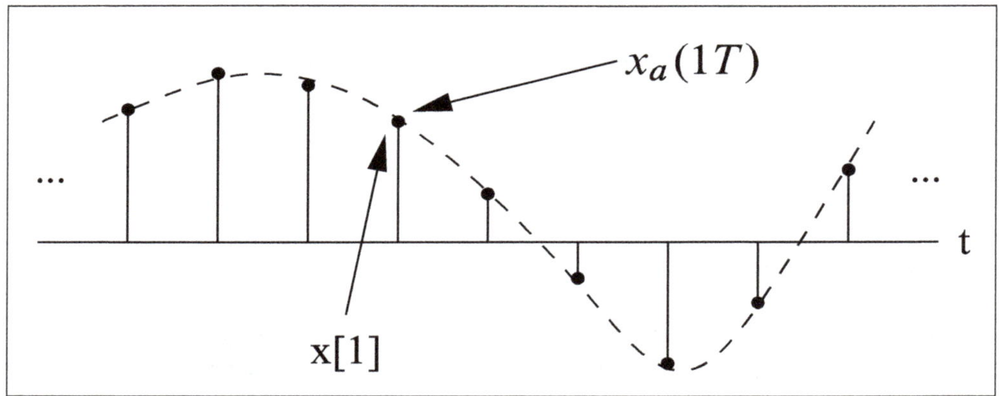

For this reason, although $x[n]$ is strictly the nth number in the sequence, we often refer to it as the nth sample. We also often refer to "the sequence $x[n]$" when we mean the entire sequence.

Discrete-time signals are often depicted graphically as follow:

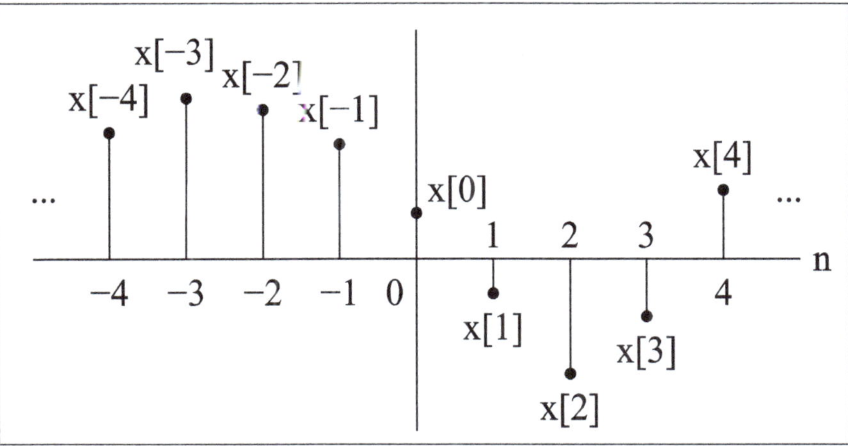

The value $x[n]$ is undefined for noninteger values of n.

Sequences can be manipulated in several ways. The sum and product of two sequences $x[n]$ and $y[n]$ are defined as the sample-by-sample sum and product respectively. Multiplication of $x[n]$ by a is defined as the multiplication of each sample value by a.

A sequence $y[n]$ is a delayed or shifted version of $x[n]$ if:

$$y[n] = x[n - n_0],$$

with n_0 an integer.

The unit sample sequence:

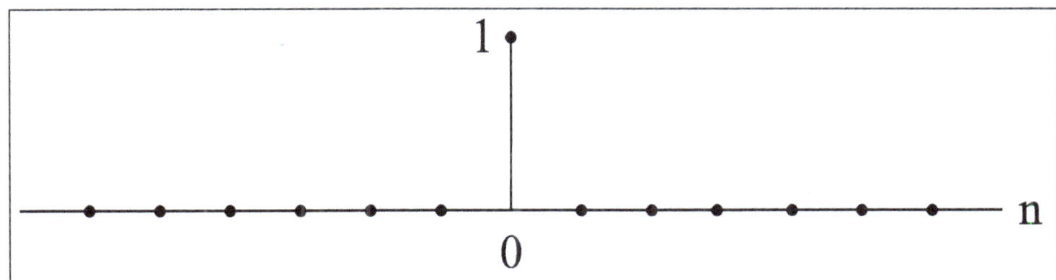

is defined as:

$$\delta[n] = \begin{cases} 0 & n \neq 0 \\ 1 & n = 0. \end{cases}$$

This sequence is often referred to as a discrete-time impulse, or just impulse. It plays the same role for discrete-time signals as the Dirac delta function does for continuous-time signals. However, there are no mathematical complications in its definition.

An important aspect of the impulse sequence is that an arbitrary sequence can be represented as a sum of scaled, delayed impulses. For example, the sequence.

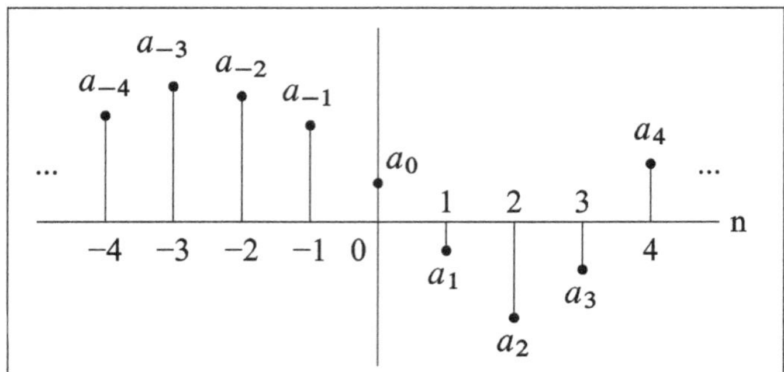

can be represented as:

$$x[n] = a_{-4}\delta[n+4] + a_{-3}\delta[n+3] + a_{-2}\delta[n+2] + a_{-1}\delta[n+1] + a_0\delta[n] \\ + a_1\delta[n-1] + a_2\delta[n-2] + a_3\delta[n-3] + a_4\delta[n-4].$$

In general, any sequence can be expressed as:

$$x[n] = \sum_{k=-\infty}^{\infty} x[k]\delta[n-k].$$

The unit step sequence:

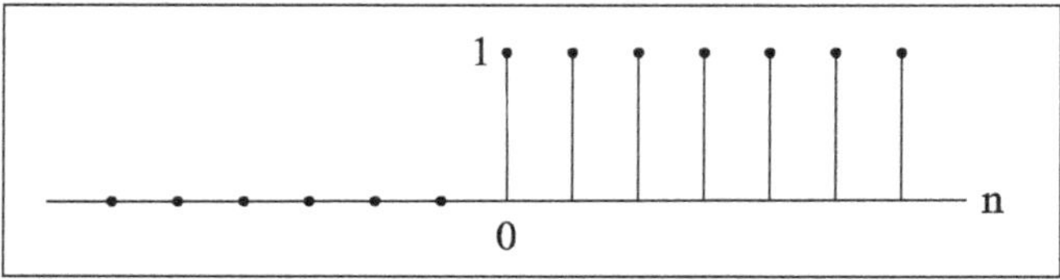

Is defined as:

$$u[n] = \begin{cases} 1 & n \geq 0 \\ 0 & n < 0. \end{cases}$$

The unit step is related to the impulse by:

$$u[n] = \sum_{k=-\infty}^{n} \delta[k].$$

Alternatively, this can be expressed as:

$$u[n] = \delta[n] + \delta[n-1] + \delta[n-2] + \cdots = \sum_{k=-\infty}^{n} \delta[n-k].$$

Conversely, the unit sample sequence can be expressed as the first backward difference of the unit step sequence:

$$\delta[n] = u[n] - u[n-1].$$

Exponential sequences are important for analysing and representing discrete-time systems. The general form is:

$$x[n] = A\alpha^n.$$

If A and α are real numbers then the sequence is real. If 0 < α < 1 and A is positive, then the sequence values are positive and decrease with increasing n:

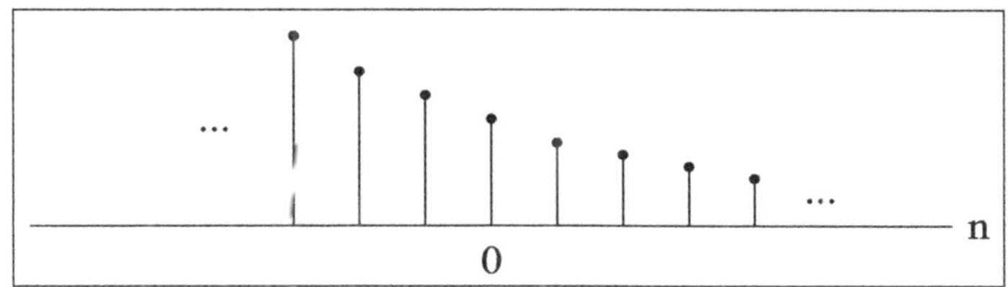

For -1 < α < 0 the sequence alternates in sign, but decreases in magnitude. For $|\alpha| > 1$ the sequence grows in magnitude as n increase:

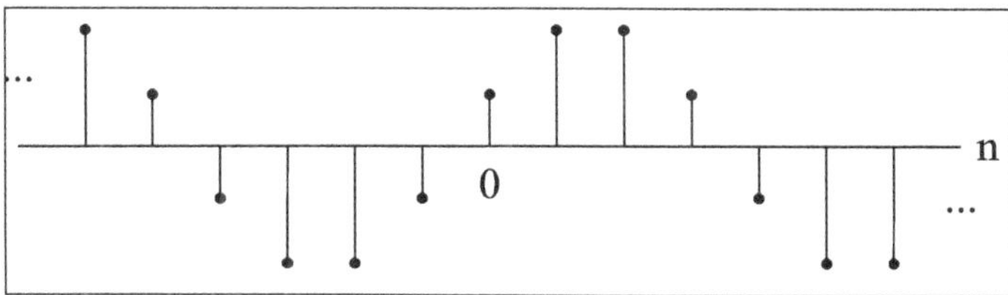

has the form:

$$x[n] = A\cos(\omega_0 n + \phi) \qquad \text{for all } n,$$

with A and ϕ real constants. The exponential sequence $A\alpha^n$ with complex $\alpha = |\alpha| e^{j\omega_0}$ and $A = |A| e^{j\phi}$ can be expressed as:

$$x[n] = A\alpha^n = |A| e^{j\phi} |\alpha|^n e^{j\omega_0 n} = |A||\alpha|^n e^{j(\omega_0 n + \phi)}$$
$$= |A||\alpha|^n \cos(\omega_0 n + \phi) + j|A||\alpha|^n \sin(\omega_0 n + \phi),$$

so the real and imaginary parts are exponentially weighted sinusoids. When $|\alpha| = 1$ the sequence is called the complex exponential sequence:

$$x[n] = |A| e^{j(\omega_0 n + \phi)} = |A| \cos(\omega_0 n + \phi) + j|A| \sin(\omega_0 n + \phi).$$

The frequency of this complex sinusoid is ω_0, and is measured in radians per sample. The phase of the signal is ϕ.

The index n is always an integer. This leads to some important differences between the properties of discrete-time and continuous-time complex exponentials:

Consider the complex exponential with frequency $(\omega_0 + 2\pi)$.

$$x[n] = Ae^{j(\omega_0 + 2\pi)} = Ae^{j\omega_0 n} e^{j2\pi n} = Ae^{j\omega_0 n}.$$

Thus the sequence for the complex exponential with frequency ω_0 is exactly the same as that for the complex exponential with frequency $(\omega_0 + 2\pi)$. More generally, complex exponential sequences with frequencies $(\omega_0 + 2\pi r)$, where r is an integer, are indistinguishable from one another. Similarly, for sinusoidal sequences:

$$x[n] = A\cos[(\omega_0 + 2\pi r)n + \phi] = A\cos(\omega_0 n + \phi).$$

In the continuous-time case, sinusoidal and complex exponential sequences are always periodic. Discrete-time sequences are periodic (with period N) if:

$$x[n] = x[n+N] \qquad \text{for all } n.$$

Thus the discrete-time sinusoid is only periodic if:

$$A\cos(\omega_0 n + \phi) = A\cos(\omega_0 n + \omega_0 N + \phi),$$

which requires that:

$$\omega_0 N = 2\pi k \quad \text{for } k \text{ an integer}.$$

The same condition is required for the complex exponential sequence $Ce^{j\omega_0 n}$ to be periodic.

The two factors just described can be combined to reach the conclusion that there are only N distinguishable frequencies for which the corresponding sequences are periodic with period N. One such set is:

$$\omega_k = \frac{2\pi k}{N}, \qquad k = 0,1,\ldots,N-1.$$

Additionally, for discrete-time sequences the interpretation of high and low frequencies has to be modified: the discrete-time sinusoidal sequence $x[n] = A\cos(\omega_0 n + \phi)$ oscillates more rapidly as ω_0 increases from 0 to π, but the oscillations become slower as it increases further from π to 2π.

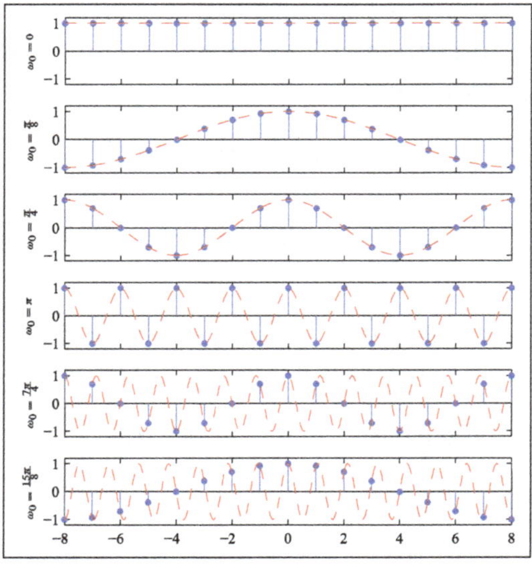

The sequence corresponding to $\omega_0 = 0$ is indistinguishable from that with $\omega_0 = 2\pi$. In general, any frequencies in the vicinity of $\omega_0 = 2\pi k$ for integer k are typically referred to as low frequencies, and those in the vicinity $\omega_0 = (\pi + 2\pi k)$ are high frequencies.

Elementary Signals

The elementary signals are used for analysis of systems. Such signals are,

- Step,
- Impulse,
- Ramp,
- Exponential,
- Sinusoidal.

Unit Step Signal

- Unit Step Sequence: The unit step signal has amplitude of 1 for positive value and amplitude of 0 for negative value of independent variable.
- It have two different parameter such as CT unit step signal u(t) and DT unit step signal u(n).
- The mathematical representation of CT unit step signal u(t).

Ramp Signal

- The amplitude of every sample is linearly increased with the positive value of independent variable.
- Mathematical representation of CT unit ramp signal is given by:

  $r(t) = t \cdot u(t)$

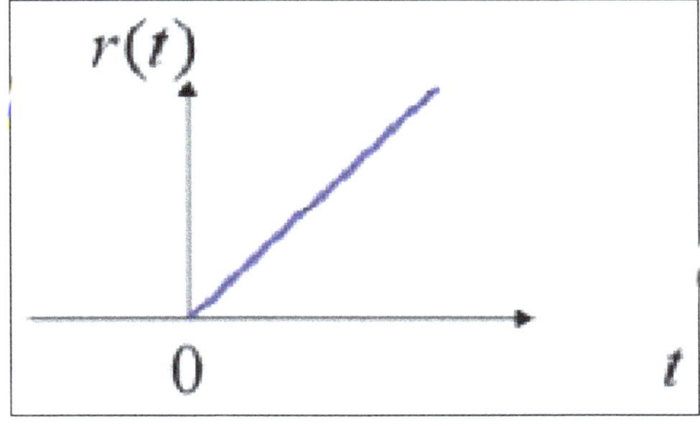

Unit Impulse Function

Amplitude of unit impulse approaches 1 as the width approaches zero and it has zero value at all other values.

The mathematical representation of unit impulse signal for CT is given by:

$$\delta(0) = \infty$$

$$\delta(0) = O, t \neq 0$$

$$\int_{-\infty}^{t} \delta(t)dt = \begin{cases} 1, & t > 0 \\ 0, & t < 0 \end{cases}$$

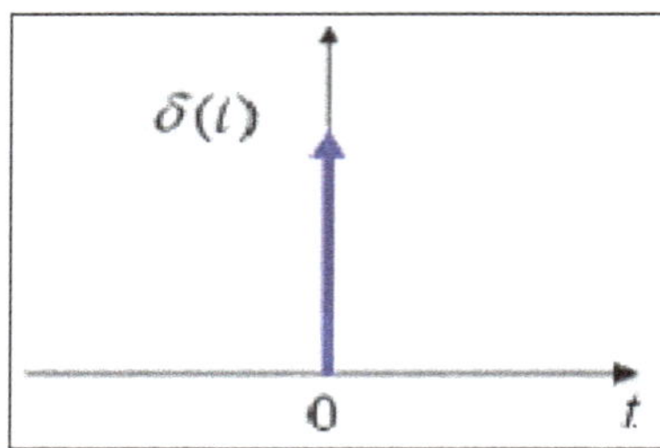

It is used to determine the impulse response of system.

Sinusoidal Signal

A continuous time sinusoidal signal is given by:

$$x(t) = A\cos\,(\Omega_0 t + \alpha)$$

Where, A – amplitute α – phase angle in radians

Exponential Signal

- It is exponentially growing or decaying signal.
- Mathematical representation for CT exponential signal is:

$$x(t) = Ce^{at}, \quad where\, C, a \in \square$$

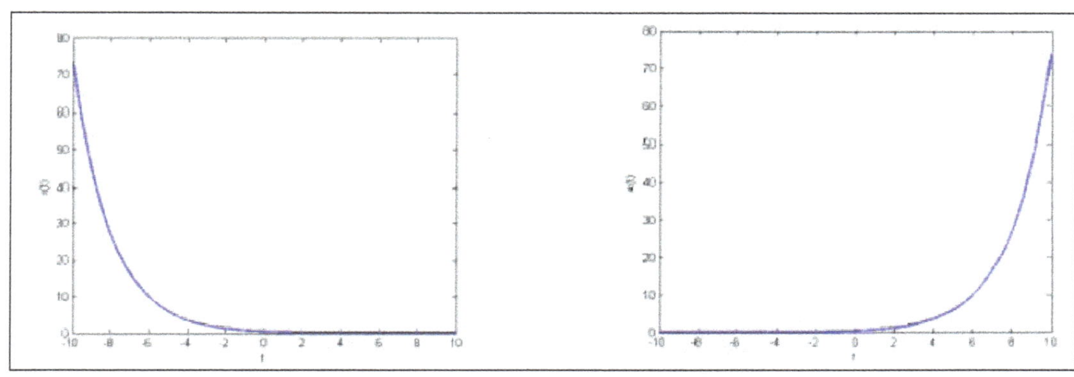

Basic Operations on Signals

A signal, comprises of a set of information expressed as a function of any number of independent variables, that can be given as an input to a system, or derived as output from the system, to realize its true practical utility. The signal we derive out of a complex system might not always be in the form we want, being well acquainted with some basic signal operations may come really handy to enhance the understandability and applicability of signals.

The mathematical transformation from one signal to another can be expressed as:

$$Y(t) = TX(t)$$

Where, $Y(t)$ represents the modified signal derived from the original signal $X(t)$, having only one independent variable t.

Time Shifting

Time shifting is, the shifting of a signal in time. This is done by adding or subtracting a quantity of the shift to the time variable in the function. Subtracting a fixed positive quantity from the time variable will shift the signal to the right (delay) by the subtracted quantity, while adding a fixed positive amount to the time variable will shift the signal to the left (advance) by the added quantity.

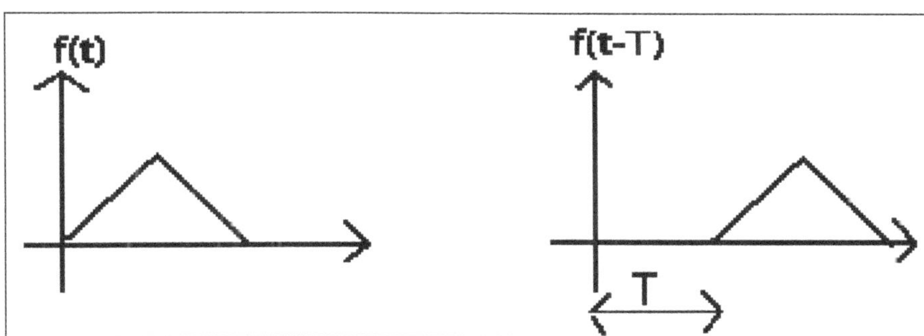

$f(t-T)$ moves (delays) f to the right by T.

Time Reversal

Whenever signal's time is multiplied by -1, it is known as time reversal of the signal. In this case, the signal produces its mirror image about Y-axis. Mathematically, this can be written as:

$$x(t) \rightarrow y(t) \rightarrow x(-t)$$

This can be best understood by the following example.

In the above example, we can clearly see that the signal has been reversed about its Y-axis. So, it is one kind of time scaling also, but here the scaling quantity is −1 always.

For any complex signal $x(n), n \in (-\infty, \infty)$, we have:

$$\text{Flip}(x) \leftrightarrow \text{Flip}(X)$$

where $\text{Flip}_n(x) \triangleq x(-n)$.

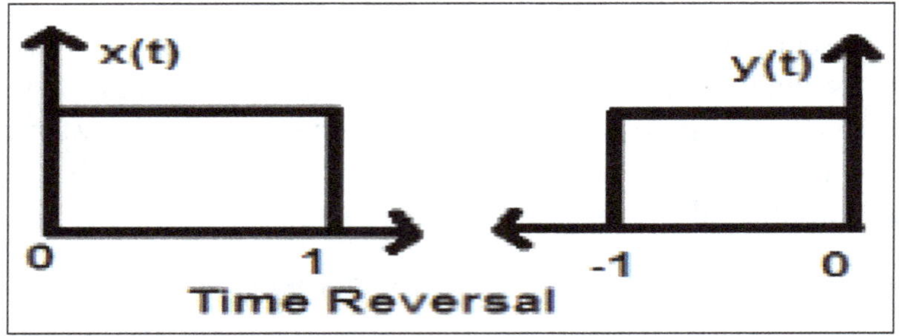

Proof:

$$\text{DTFT}_\omega(\text{Flip}(x)) \triangleq \sum_{n=-\infty}^{\infty} x(-n)e^{-j\omega n} = \sum_{m=\infty}^{-\infty} x(m)e^{-j(-\omega)m} X(-\omega)$$

$$\triangleq \text{Flip}_\omega(X)$$

Arguably, $\text{Flip}(x)$ should include complex conjugation. Let:

$$\text{Flip}'_n(x) \triangleq \overline{\text{FLIP}_n(x)} = \overline{x(-n)}$$

denote such a definition. Then in this case we have:

$$\text{Flip}'(x) \leftrightarrow \overline{\text{X}}$$

Proof:

$$\text{DTFT}_\omega(\text{Flip}'(x)) \triangleq \sum_{n=-\infty}^{\infty} \overline{x(-n)}e^{-j\omega n} = \sum_{m=\infty}^{-\infty} \overline{x(m)e^{-j\omega m}} \triangleq \overline{X(\omega)}$$

In the typical special case of real signals $(x(n) \in \mathbf{R})$, we have $\text{Flip}(x) = \text{Flip}'(x)$ so that:

$$\text{Flip}(x) \leftrightarrow \overline{X}.$$

Amplitude Scaling

Amplitude scaling means changing an amplitude of given continuous time signal. We will denote continuous time signal by $x(t)$. If it is multiplied by some constant 'B' then resulting signal is:

$$y(t) = B\,x(t)$$

Example: Sketch $y(t) = 5u(t)$

Solution: we know that $u(t)$ is unit step function. So if we multiply it with 5, its amplitude will become 5 and it shown as follows:

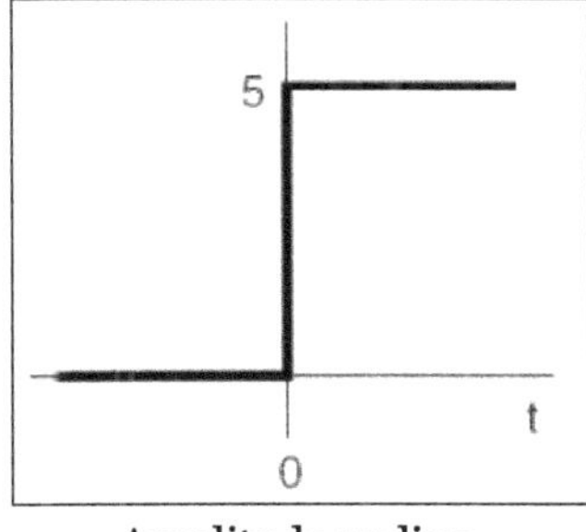

Amplitude scaling.

Time Scaling

Time scaling compresses or dilates a signal by multiplying the time variable by some quantity. If that quantity is greater than one, the signal becomes narrower and the operation is called compression, while if the quantity is less than one, the signal becomes wider and is called dilation.

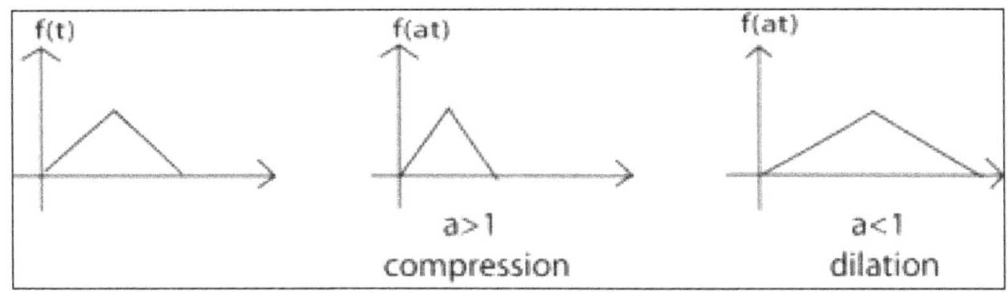

$f(at)$ compresses f by a.

Example:

Given $f(t)$ we would like to plot $f(at-b)$. The figure below describes a method to accomplish this.

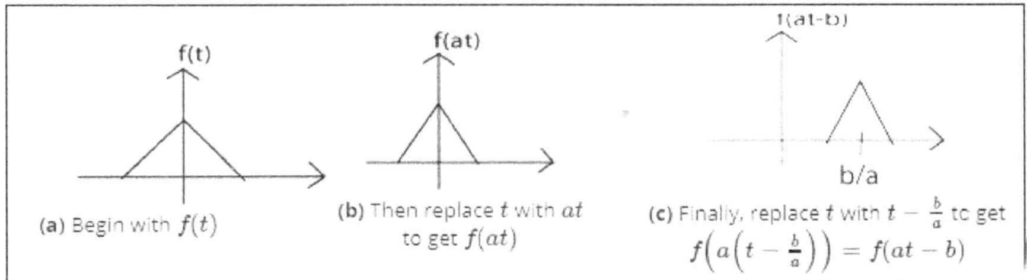

Basic Operations in Signal Processing

Addition and Subtraction of Signals

The first and foremost operation which we will consider will be addition. The addition of signals is very similar to traditional mathematics. That is, if $x_1(t)$ and $x_2(t)$ are the two continuous time signals, then the addition of these two signals is expressed as $x_1(t)+x_2(t)$.

The resultant signal can be represented as $y(t)$ from which we can write:

$$y(t) = x_1(t) + x_2(t)$$

Similarly for discrete time signals, $x_1(t)$ and $x_2(t)$, we can write:

$$y[n] = x_1[n] + x_2[n]$$

Figure shows an example of addition operation performed over the continuous time signals $x_1(t)$ and $x_2(t)$.

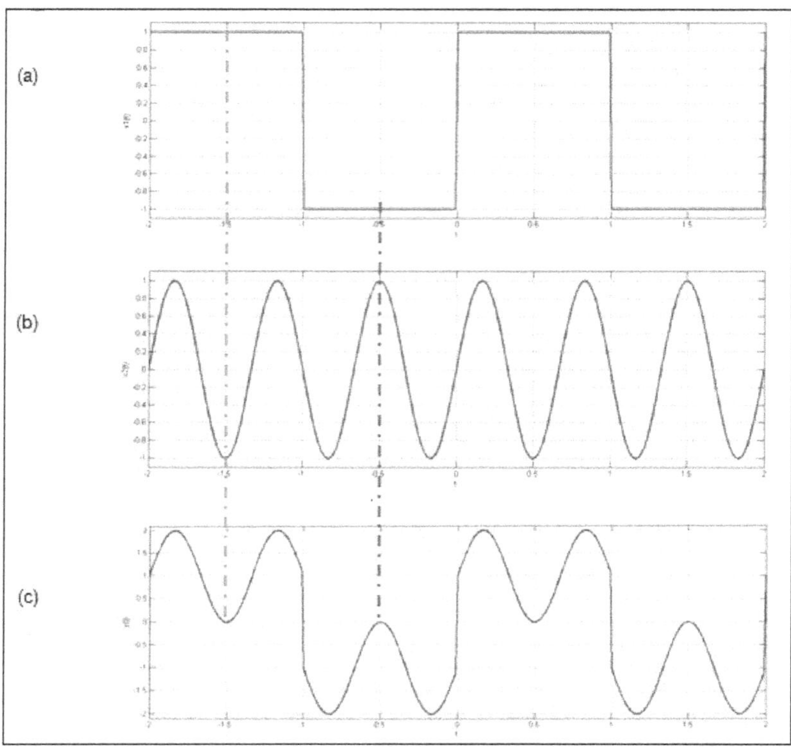

Addition operation performed on two continuous time signals.

By following the green-coloured dotted line in figure, you can note the value of $y(t)$ at. $t = -1.5$ to be 0 which is nothing but the summation of $x_1(t)$ at $t = -1.5$ which is 1 and that of $x_2(t)$ at $t = -1.5$ which is -1. Similarly, by moving along the purple-coloured dotted line, the value of $y(-0.5)$ is seen to be 0 which is equal to $x_1(-0.5) + x_2(-0.5) = -1 + 1$.

Hence it can be concluded that all the values of the resultant signal $y(t)$ can be obtained by adding the corresponding values of the signals $x_1(t)$ and $x_2(t)$. Although we have depicted the example of continuous time signals, the conclusion stated holds good even for discrete time signals.

Practical Scenario

A practical aspect in which signal addition plays its role is in the case of transmission of a signal through a communication channel. This is because, here, we see that the undesired noise gets added up with the desired signal.

Another example which can be quoted is of dithering where the noise is added to the signal intentionally. This is because, when done so, one can effectively reduce undesired artifacts created as an aftermath of quantization errors.

Subtraction

Similar to the case of addition, subtraction deals with the subtraction of two or more signals in order to obtain a new signal. Mathematically it can be represented as:

$$y(t) = x_1(t) - x_2(t), \text{for continuous time signals}, x_1(t) \text{ and } x_2(t)$$

and

$$y[n] = x_1[n] - x_2[n] \text{ , for discrete time signals}, x_1[n] \text{ and } x_2[n]$$

Subtraction operation performed over two discrete time signals $x_1[n]$ and $x_2[n]$.

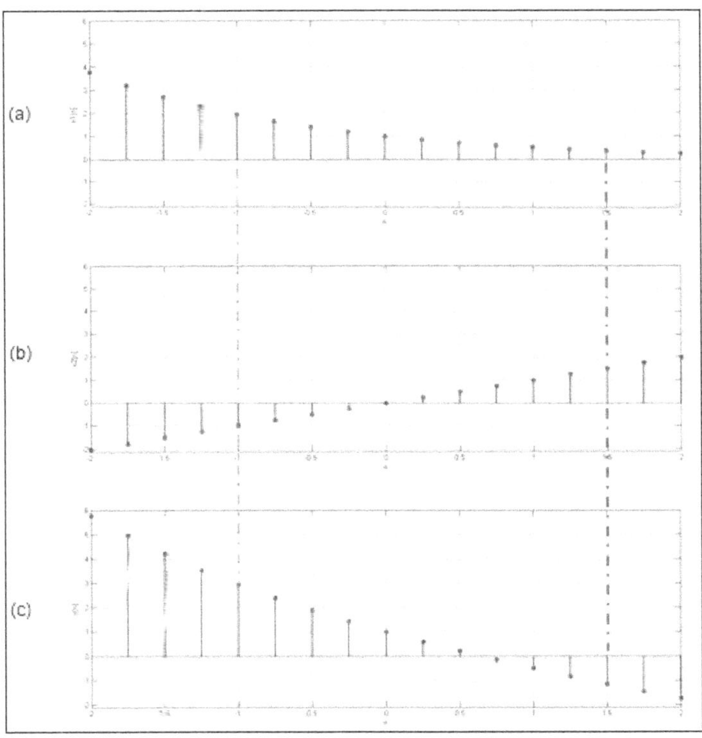

Subtraction operation performed on two discrete time signals.

Even in the case of subtraction operation, all the values of the resultant signal y[n] can be obtained by subtracting the corresponding values of the signals x1[n] and x2[n].

This is evident from the figure as the discontinuous green-colored dotted line shows $y[-1] = 3$ which is equal to $x_1[-1] - x_2[-1] = 2 - (-1)$. Another example of a similar kind is shown by the discontinuous purple-colored dotted line, wherein $y[1.5] = x_1[1.5] - x_2[1.5] = 0.4 - 1.5 = -1.1$.

It can be stated that the conclusion we arrive at in the case of subtraction operation is very similar to that of the addition operation and applies to both continuous and discrete time signals.

Practical Scenario

One practical aspect which connects with that of subtracting the signals is that of a Moving Target Indicator (MTI) used in radar communications. Here the most recent signal is subtracted from its previous version so as to obtain the signal which indicates just the moving targets by eliminating the stationary ones. This is very much necessary so as to facilitate PPI (Plan Position Indicator) display of radar systems.

Yet another example which extensively makes use of signal subtraction is the design of closed-loop control systems. Such systems employ negative feedback in order to accurately control an output variable, and this negative-feedback structure relies upon subtraction (the feedback signal is subtracted from the setpoint signal).

Multiplication, Differentiation and Integration of Signals

The next basic signal operation performed over the dependent variable is multiplication. In this case, as you might have already guessed, two or more signals will be multiplied so as to obtain the new signal.

Mathematically, this can be given as:

$$y(t) = x_1(t) \times x_2(t)\text{, for continuous} - \text{time signals } x_1(t) \textit{ and } x_2(t)$$

and

$$y[n] = x_1[n] \times x_2[n]\text{, for discrete} - \text{time signals } x_1[n] \text{ and } x_2[n]$$

The resultant discrete-time signal $y[n]$ obtained by multiplying the two discrete-time signals $x_1[n]$ and $x_2[n]$ shown in figures (a) and 1(b), respectively.

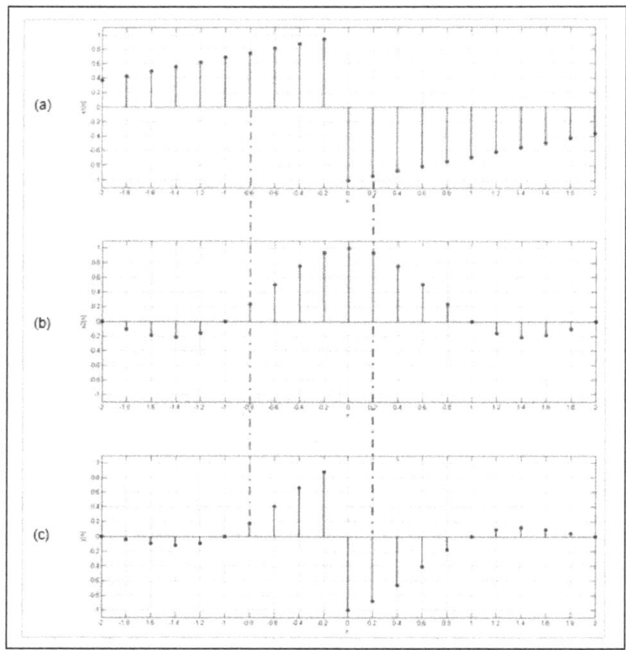

Multiplication operation performed over two discrete-time signals.

Here the value of $y[n]$ at $n=-0.8$ is seen to be 0.17, which is found to be equal to the product of the values of $x_1[n]$ and $x_2[n]$ at $n=-0.8$, which are 0.75 and 0.23, respectively. In other words, by tracing along the green dotted-dashed line, one gets 0.75 × 0.23 = 0.17.

Similarly, if we move along the purple dotted-dashed line (at $n=0.2$) to collect the values of $x_1[n], x_2[n]$, and $y[n]$, we find that they are -0.94, 0.94, and -0.88, respectively. Here also we find that -0.94 × 0.94 = -0.88, which in turn implies $x_1[0.2] \times x_2[0.2] = y[0.2]$.

Thus, we can conclude that the multiplication operation results in the generation of a signal whose values can be obtained by multiplying the corresponding values of the original signals. This is true irrespective of whether we are dealing with a continuous-time or discrete-time signal.

Practical Scenario

Multiplication of signals is exploited in the field of analog communication when performing amplitude modulation (AM). In AM, the message signal is multiplied with the carrier signal so as to obtain a modulated signal.

Another example in which signal multiplication plays an important role is frequency shifting in RF (radio frequency) systems. Frequency shifting is a fundamental aspect of RF communication, and it is accomplished using a mixer, which is similar to an analog multiplier.

Differentiation

The next signal operation which is important in signal processing is differentiation. A signal is differentiated to determine the rate at which it changes. That is, if x(t) is the continuous-time signal, then its differentiation yields the output signal $y(t)$, given by $y(t)=\frac{d}{dt}\{x(t)\}$.

Figure shows an example of a signal along with its differentiation. The figure shows the first derivative of a parabola—in figure (a)—spanning from t = 0 to 2 to be a ramp—in figure (b)—which has its values ranging from 0 to 4. The first derivative of the ramp in figure (a) spanning from $t=2$ to 6 is shown to be a constant amplitude of 1 in figure (b).

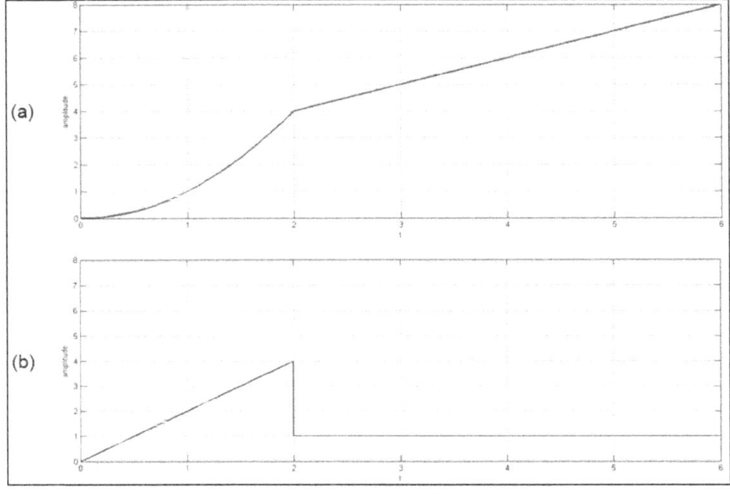

An original signal and its differentiation.

Next, you should note that the differentiation operation is not restricted to continuous-time signals; it is also applicable to discrete-time signals.

Also, keep in mind that a signal can be differentiated more than once. For example, differentiating an original signal leads to a "first derivative" and differentiating this first derivative produces the "second derivative".

Practical Scenario

Differentiation of a signal takes the form of the gradient operator in the field of image or video processing. In the case of image processing, the gradient technique is a popular method which is used to detect the edges in the given image. With video processing, this operator is used for motion detection. This kind of processing is important in the field of robotics.

In addition, many control and tracking applications, such as in aeronautical systems, make use of real-time differentiators. This is because these applications require highly accurate data pertaining to velocity and acceleration. By using differentiators, this data can be obtained directly from position sensors, reducing the need for other sensors.

Integration

Integration is the counterpart of differentiation. If we integrate a signal $x(t)$, the result $y(t)$ is represented as $\int x(t)$. Graphically, the act of integration computes the area under the curve of the original signal.

In figure, a composite signal composed of a ramp extending from $t=0$ to 2 and a constant value ranging from $t=2$ to 5 is being integrated. The output obtained is shown in figure (b); the integration of the ramp has resulted in a parabola (extending from $t=0$ to 2), and the integration of the constant value has created a ramp (ranging from $t=2$ to 5).

As with differentiation, we can integrate a signal multiple times.

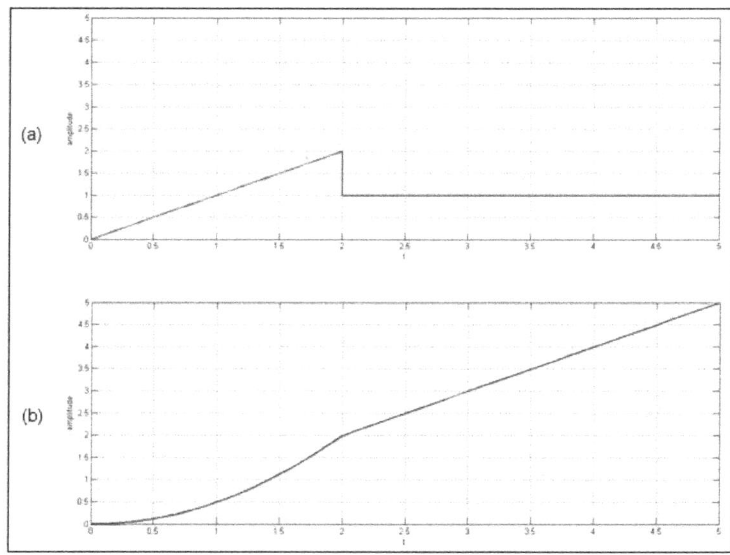

The integration operation.

Practical Scenario

Integration is fundamental in signal-processing operations such as the Fourier transform, correlation, and convolution. These are, in turn, used to analyze different properties of a signal.

Other applications that employ integration are those in which small input currents are converted, via integration, into larger output voltages. Charge amplifiers are used with piezoelectric sensors, photodiodes, and CCD imagers. Also, charge amplifiers can be used to convert an accelerometer output into velocity and displacement signals, because integrating acceleration yields velocity, and integrating velocity yields displacement.

References

- Introduction-to-signals-types-properties-operation-application: electricaltechnology.org, Retrieved 13 May, 2019
- Analog-signal, computernetworkingnotes-communication-networks: ecomputernotes.com, Retrieved 18 August, 2019
- Analog-digital-conversion: geeksforgeeks.org, Retrieved 31 July, 2019
- Elementary-Signals-13342: brainkart.com, Retrieved 25 April, 2019
- Basic-signal-operations: electrical4u.com, Retrieved 05 January, 2019
- Basic-operations-in-signals-overview, technical-articles: allaboutcircuits.com, Retrieved 25 June, 2019
- Operations-signal-processing-multiplication-differentiation-integration, technical-articles: allaboutcircuits.com, Retrieved 19 May, 2019

2

Signal Processing and Quantization

Quantization refers to the process of mapping infinite input values from a large set to a finite set of output values. Digital signal processing, analog signal processing, Nyquist–Shannon sampling theorem, aliasing, etc. fall under the domain of signal processing. This chapter has been carefully written to provide an easy understanding of signal processing and quantization.

Signal Processing

The processing of signals by means of hardwired or programmable devices, the signals being regarded as continuous or discrete and being approximated by analog or digital devices accordingly. Filtering and image processing are examples of signal processing.

Digital Signal Processing

Digital signal processing (DSP) is the use of digital processing, such as by computers or more specialized digital signal processors, to perform a wide variety of signal processing operations. The signals processed in this manner are a sequence of numbers that represent samples of a continuous variable in a domain such as time, space, or frequency.

Digital signal processing and analog signal processing are subfields of signal processing. DSP applications include audio and speech processing, sonar, radar and other sensor array processing, spectral density estimation, statistical signal processing, digital image processing, signal processing for telecommunications, control systems, biomedical engineering, seismology, among others.

DSP can involve linear or nonlinear operations. Nonlinear signal processing is closely related to nonlinear system identification and can be implemented in the time, frequency, and spatio-temporal domains.

The application of digital computation to signal processing allows for many advantages over analog processing in many applications, such as error detection and correction in transmission as well as data compression. DSP is applicable to both streaming data and static (stored) data.

Domains

In DSP, engineers usually study digital signals in one of the following domains: time domain (one-dimensional signals), spatial domain (multidimensional signals), frequency domain, and wavelet domains. They choose the domain in which to process a signal by making an informed assumption (or by trying different possibilities) as to which domain best represents the essential characteristics of the signal and the processing to be applied to it. A sequence of samples from a measuring device produces a temporal or spatial domain representation, whereas a discrete Fourier transform produces the frequency domain representation.

Time and Space Domains

The most common processing approach in the time or space domain is enhancement of the input signal through a method called filtering. Digital filtering generally consists of some linear transformation of a number of surrounding samples around the current sample of the input or output signal. There are various ways to characterize filters; for example:

- A linear filter is a linear transformation of input samples; other filters are nonlinear. Linear filters satisfy the superposition principle, i.e. if an input is a weighted linear combination of different signals, the output is a similarly weighted linear combination of the corresponding output signals.
- A causal filter uses only previous samples of the input or output signals; while a non-causal filter uses future input samples. A non-causal filter can usually be changed into a causal filter by adding a delay to it.
- A time-invariant filter has constant properties over time; other filters such as adaptive filters change in time.
- A stable filter produces an output that converges to a constant value with time, or remains bounded within a finite interval. An unstable filter can produce an output that grows without bounds, with bounded or even zero input.
- A finite impulse response (FIR) filter uses only the input signals, while an infinite impulse response (IIR) filter uses both the input signal and previous samples of the output signal. FIR filters are always stable, while IIR filters may be unstable.

A filter can be represented by a block diagram, which can then be used to derive a sample processing algorithm to implement the filter with hardware instructions. A filter may also be described as a difference equation, a collection of zeros and poles or an impulse response or step response.

The output of a linear digital filter to any given input may be calculated by convolving the input signal with the impulse response.

Frequency Domain

Signals are converted from time or space domain to the frequency domain usually through use of the Fourier transform. The Fourier transform converts the time or space information to a magnitude and phase component of each frequency. With some applications, how the phase varies with frequency can be a significant consideration. Where phase is unimportant, often the Fourier transform is converted to the power spectrum, which is the magnitude of each frequency component squared.

The most common purpose for analysis of signals in the frequency domain is analysis of signal properties. The engineer can study the spectrum to determine which frequencies are present in the input signal and which are missing. Frequency domain analysis is also called *spectrum-* or *spectral analysis*.

Filtering, particularly in non-realtime work can also be achieved in the frequency domain, applying the filter and then converting back to the time domain. This can be an efficient implementation and can give essentially any filter response including excellent approximations to brickwall filters.

There are some commonly-used frequency domain transformations. For example, the cepstrum converts a signal to the frequency domain through Fourier transform, takes the logarithm, then applies another Fourier transform. This emphasizes the harmonic structure of the original spectrum.

Z-plane Analysis

Digital filters come in both IIR and FIR types. Whereas FIR filters are always stable, IIR filters have feedback loops that may become unstable and oscillate. The Z-transform provides a tool for analyzing stability issues of digital IIR filters. It is analogous to the Laplace transform, which is used to design and analyze analog IIR filters.

Wavelet

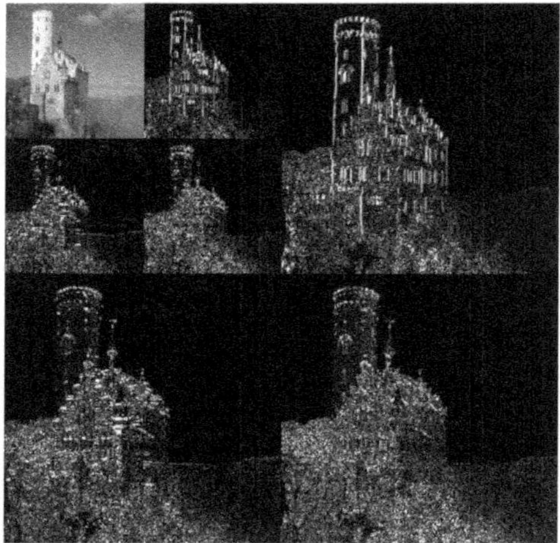

An example of the 2D discrete wavelet transform that is used in JPEG2000. The original image is high-pass filtered, yielding the three large images, each describing local changes in brightness (details) in the original image. It is then low-pass filtered and downscaled, yielding an approximation image; this image is high-pass filtered to produce the three smaller detail images, and low-pass filtered to produce the final approximation image in the upper-left.

In numerical analysis and functional analysis, a discrete wavelet transform (DWT) is any wavelet transform for which the wavelets are discretely sampled. As with other wavelet transforms, a key advantage it has over Fourier transforms is temporal resolution: it captures both frequency *and* location information.The accuracy of the joint time-frequency resolution is limited by the uncertainty principle of time-frequency.

Applications

Applications of DSP include audio signal processing, audio compression, digital image processing, video compression, speech processing, speech recognition, digital communications, digital synthesizers, radar, sonar, financial signal processing, seismology and biomedicine. Specific examples include speech coding and transmission in digital mobile phones, room correction of sound in hi-fi and sound reinforcement applications, weather forecasting, economic forecasting, seismic data processing, analysis and control of industrial processes, medical imaging such as CAT scans and MRI, MP3 compression, computer graphics, image manipulation, audio crossovers and equalization, and audio effects units.

Implementation

DSP algorithms may be run on general-purpose computers and digital signal processors. DSP algorithms are also implemented on purpose-built hardware such as application-specific integrated circuit (ASICs). Additional technologies for digital signal processing include more powerful general purpose microprocessors, field-programmable gate arrays (FPGAs), digital signal controllers (mostly for industrial applications such as motor control), and stream processors.

For systems that do not have a real-time computing requirement and the signal data (either input or output) exists in data files, processing may be done economically with a general-purpose computer. This is essentially no different from any other data processing, except DSP mathematical techniques (such as the FFT) are used, and the sampled data is usually assumed to be uniformly sampled in time or space. An example of such an application is processing digital photographs with software such as photoshop.

When the application requirement is real-time, DSP is often implemented using specialized or dedicated processors or microprocessors, sometimes using multiple processors or multiple processing cores. These may process data using fixed-point arithmetic or floating point. For more demanding applications FPGAs may be used. For the most demanding applications or high-volume products, ASICs might be designed specifically for the application.

Downsampling

In digital signal processing, downsampling and decimation are terms associated with the process of resampling in a multi-rate digital signal processing system. Both terms are used by various authors to describe the entire process, which includes lowpass filtering, or just the part of the process that does not include filtering. When downsampling (decimation) is performed on a sequence of samples of a *signal* or other continuous function, it produces an approximation of the sequence that would have been obtained by sampling the signal at a lower rate (or density, as in the case of a photograph). The *decimation factor* is usually an integer or a rational fraction greater than one. This factor multiplies the sampling interval or, equivalently, divides the sampling rate. For example, if compact disc audio at 44,100 samples/second is decimated by a factor of 5/4, the resulting sample rate is 35,280. A system component that performs decimation is called a *decimator*.

Downsampling by an Integer Factor

Rate reduction by an integer factor M can be explained as a two-step process, with an equivalent implementation that is more efficient:

- Reduce high-frequency signal components with a digital lowpass filter.
- *Decimate* the filtered signal by M; that is, keep only every M^{th} sample. A notation for this operation is: $x[Mn] = x[n]_{\downarrow M}$.

Step 2 alone allows high-frequency signal components to be misinterpreted by subsequent users of the data, which is a form of distortion called aliasing. Step 1, when necessary, suppresses aliasing to an acceptable level. In this application, the filter is called an anti-aliasing filter,

When the anti-aliasing filter is an IIR design, it relies on feedback from output to input, prior to the second step. With FIR filtering, it is an easy matter to compute only every M^{th} output. The calculation performed by a decimating FIR filter for the nth output sample is a dot product:

$$y[n] = \sum_{k=0}^{K-1} x[nM-k] \cdot h[k],$$

where the $h[\bullet]$ sequence is the impulse response, and K is its length. $x[\bullet]$ represents the input sequence being downsampled. In a general purpose processor, after computing $y[n]$, the easiest way to compute $y[n+1]$ is to advance the starting index in the $x[\bullet]$ array by M, and recompute the dot product. In the case M=2, $h[\bullet]$ can be designed as a half-band filter, where almost half of the coefficients are zero and need not be included in the dot products.

Impulse response coefficients taken at intervals of M form a subsequence, and there are M such subsequences (phases) multiplexed together. The dot product is the sum of the dot products of each subsequence with the corresponding samples of the $x[\bullet]$ sequence. Furthermore, because of downsampling by M, the stream of $x[\bullet]$ samples involved in any one of the M dot products is never involved in the other dot products. Thus M low-order FIR filters are each filtering one of M multiplexed *phases* of the input stream, and the M outputs are being summed. This viewpoint offers a different implementation that might be advantageous in a multi-processor architecture. In other words, the input stream is demultiplexed and sent through a bank of M filters whose outputs are summed. When implemented that way, it is called a polyphase filter.

For completeness, we now mention that a possible, but unlikely, implementation of each phase is to replace the coefficients of the other phases with zeros in a copy of the $h[\bullet]$ array, process the original $x[\bullet]$ sequence at the input rate, and decimate the output by a factor of M. The equivalence of this inefficient method and the implementation is known as the *first Noble identity*.

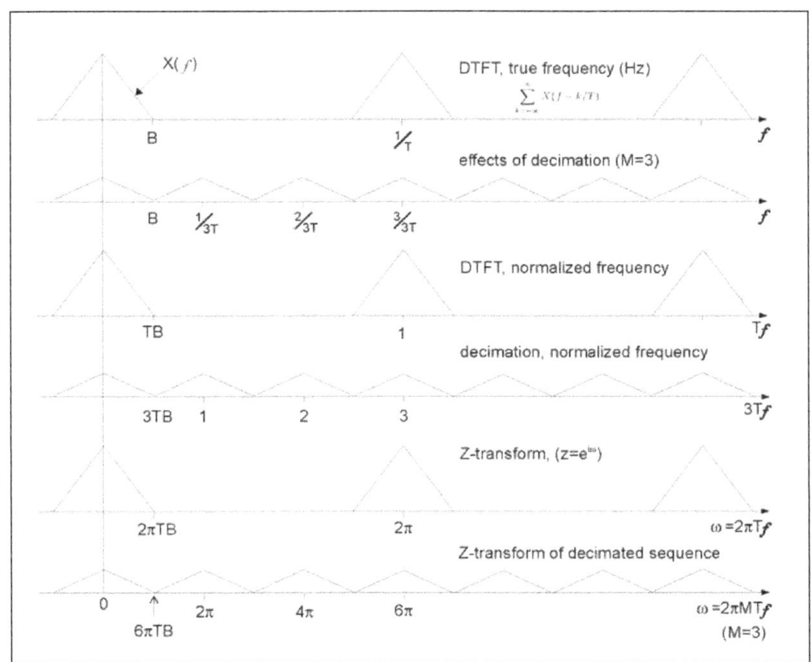

Spectral effects of decimation compared on 3 popular frequency scale conventions.

The requirements of the anti-aliasing filter can be deduced from any of the three pairs of graphs in figure. Note that all three pairs are identical, except for the units of the abscissa variables. The upper graph of each pair is an example of the periodic frequency distribution of a sampled function, $x(t)$, with Fourier transform, $X(f)$. The lower graph is the new distribution that results when $x(t)$ is sampled three times slower, or (equivalently) when the original sample sequence is decimated by a factor of M=3. In all three cases, the condition that ensures the copies of $X(f)$ do not overlap each other is the same: $B < \frac{1}{M} \cdot \frac{1}{2T}$, where T is the interval between samples, $1/T$ is the sample-rate, and $1/(2T)$ is the Nyquist frequency. The anti-aliasing filter that can ensure the condition is met has a cutoff frequency less than $\frac{1}{M}$ times the Nyquist frequency.

The abscissa of the top pair of graphs represents the discrete-time Fourier transform (DTFT), which is a Fourier series representation of a periodic summation of $X(f)$:

$$\underbrace{\sum_{n=-\infty}^{\infty} \overbrace{x(nT)}^{x[n]} \mathrm{e}^{-i2\pi fnT}}_{\text{DTFT}} = \frac{1}{T} \sum_{k=-\infty}^{\infty} X\left(f - \frac{k}{T}\right).$$

When T has units of seconds, f has units of hertz. Replacing T with MT in the formulas above gives the DTFT of the decimated sequence, $x[nM]$:

$$\sum_{n=-\infty}^{\infty} x(n \cdot MT)\, \mathrm{e}^{-i2\pi fn(MT)} = \frac{1}{MT} \sum_{k=-\infty}^{\infty} X\left(f - \tfrac{k}{MT}\right).$$

The periodic summation has been reduced in amplitude and periodicity by a factor of M, as depicted in the second graph of figure. Aliasing occurs when adjacent copies of $X(f)$ overlap.

The purpose of the anti-aliasing filter is to ensure that the reduced periodicity does not create overlap.

In the middle pair of graphs, the frequency variable, f has been replaced by normalized frequency, which creates a periodicity of 1 and a Nyquist frequency of ½. A common practice in filter design programs is to assume those values and request only the corresponding cutoff frequency in the same units. In other words, the cutoff frequency $B_{max} = \frac{1}{M} \cdot \frac{1}{2T}$, is normalized to $TB_{max} = \frac{1}{M} \cdot \frac{1}{2} = \frac{0.5}{M}$. The units of this quantity are (seconds/sample)×(cycles/second) = cycles/sample.

The bottom pair of graphs represent the Z-transforms of the original sequence and the decimated sequence, constrained to values of complex-variable, z, of the form $z = e^{i\omega}$. Then the transform of the $x[n]$ sequence has the form of a Fourier series. By comparison with Eq.1, we deduce:

$$\sum_{n=-\infty}^{\infty} x[n]\, z^{-n} = \sum_{n=-\infty}^{\infty} x(nT)\, e^{-i\omega n} = \frac{1}{T} \sum_{k=-\infty}^{\infty} \underbrace{X\left(\tfrac{\omega}{2\pi T} - \tfrac{k}{T}\right)}_{X\left(\frac{\omega - 2\pi k}{2\pi T}\right)},$$

which is depicted by the fifth graph in figure. Similarly, the sixth graph depicts:

$$\sum_{n=-\infty}^{\infty} x[nM]\, z^{-n} = \sum_{n=-\infty}^{\infty} x(nMT)\, e^{-i\omega n} = \frac{1}{MT} \sum_{k=-\infty}^{\infty} \underbrace{X\left(\tfrac{\omega}{2\pi MT} - \tfrac{k}{MT}\right)}_{X\left(\frac{\omega - 2\pi k}{2\pi MT}\right)}.$$

By a Rational Factor

Let M/L denote the decimation factor, where: M, L ∈ Z; M > L.

- Increase (resample) the sequence by a factor of L. This is called Upsampling, or interpolation.
- Decimate by a factor of M.

Step 1 requires a lowpass filter after increasing (*expanding*) the data rate, and step 2 requires a lowpass filter before decimation. Therefore, both operations can be accomplished by a single filter with the lower of the two cutoff frequencies. For the M > L case, the anti-aliasing filter cutoff, $\frac{0.5}{M}$ *cycles per intermediate sample*, is the lower frequency.

By an Irrational Factor

Techniques for decimation (and sample-rate conversion in general) by factor R ∈ R^+ include polynomial interpolation and the farrow structure.

Combined Methods of Decimation

An important factor in the development of digital antenna arrays for radars and Massive MIMO is the need to reduce the cost per channel. Combining the decimation process not only with an anti-aliasing filter, but also with the digital frequency shifting and I/Q-demodulation as well can help to bring down this cost.

In the simpler case of decimation of OFDM signals by an integer factor M, the algorithm may be used:

$$y[n] = \sum_{k=0}^{M-1} x[nM+k]\, e^{-i2\pi fkT}, n = 0,1,.., N$$
,

where T is interval between samples of signal and f is the central carrier frequency of the OFDM signal.

This algorithm is only one filter of the full discrete Fourier transform and can be useful to decimate samples in an ADC before digital beamforming in digital antenna arrays.

If more effective anti-aliasing filtering is required then this method may be modified to produce:

$$y[n] = \sum_{k=0}^{M-1} x[nM+k]h[k]\, e^{-i2\pi fkT}, n = 0,1,.., N$$
.

Upsampling

In digital signal processing, upsampling, expansion, and interpolation are terms associated with the process of resampling in a multi-rate digital signal processing system. *Upsampling* can be synonymous with *expansion*, or it can describe an entire process of *expansion* and filtering (*interpolation*). When upsampling is performed on a sequence of samples of a *signal* or other continuous function, it produces an approximation of the sequence that would have been obtained by sampling the signal at a higher rate (or density, as in the case of a photograph). For example, if compact disc audio at 44,100 samples/second is upsampled by a factor of 5/4, the resulting sample-rate is 55,125.

Upsampling by an Integer Factor

Rate increase by an integer factor L can be explained as a 2-step process, with an equivalent implementation that is more efficient:

- Expansion: Create a sequence, $x_L[n]$, comprising the original samples, $x[n]$, separated by $L-1$ zeros. A notation for this operation is: $x_L[n] = x(n)_{\uparrow L}$.
- Interpolation: Smooth out the discontinuities with a lowpass filter, which replaces the zeros.

In this application, the filter is called an interpolation filter, When the interpolation filter is an FIR type, its efficiency can be improved, because the zeros contribute nothing to its dot product calculations. It is an easy matter to omit them from both the data stream and the calculations. The calculation performed by an efficient interpolating FIR filter for each output sample is a dot product:

$$y[j+nL] = \sum_{k=0}^{K} x[n-k] \cdot h[j+kL], \quad j = 0,1,\ldots,L-1,$$

where the $h[\bullet]$ sequence is the impulse response, and K is the largest value of k for which $h[j + kL]$ is non-zero. In the case $L = 2$, $h[\bullet]$ can be designed as a half-band filter, where almost half of the coefficients are zero and need not be included in the dot products. Impulse response coefficients taken at intervals of L form a subsequence, and there are L such subsequences (called phases) multiplexed together. Each of L phases of the impulse response is filtering the same sequential values of the $x[\bullet]$ data stream and producing one of L sequential output values. In some multi-processor architectures, these dot products are performed simultaneously, in which case it is called a polyphase filter.

For completeness, we now mention that a possible, but unlikely, implementation of each phase is to replace the coefficients of the other phases with zeros in a copy of the $h[\bullet]$ array, and process the $x_L[n]$ sequence at L times faster than the original input rate. Then L-1 of every L outputs are zero. The desired $y[\bullet]$ sequence is the sum of the phases, where L-1 terms of the each sum are identically zero. Computing L-1 zeros between the useful outputs of a phase and adding them to a sum is effectively decimation. It's the same result as not computing them at all. That equivalence is known as the *second Noble identity.*

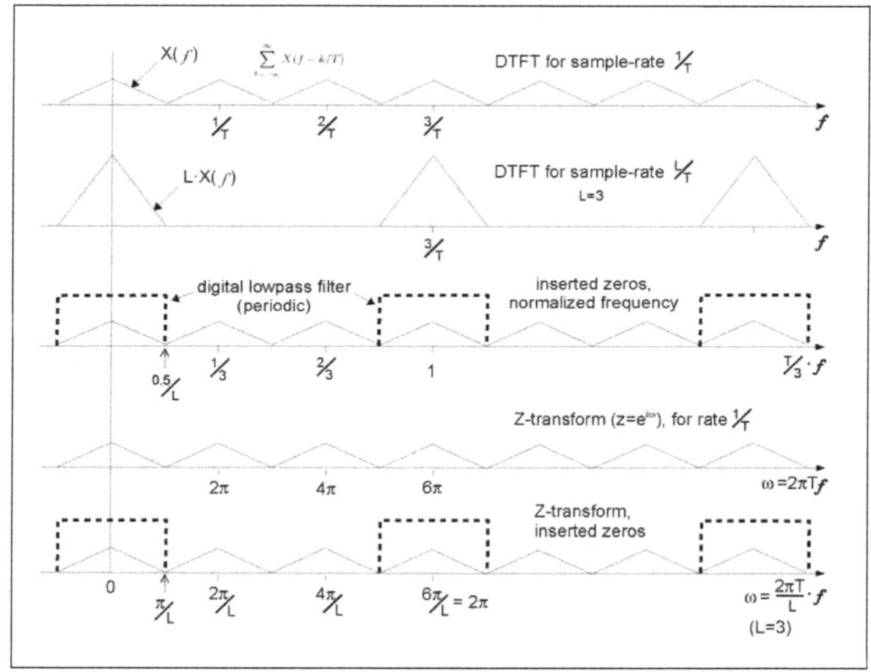

Spectral depictions of zero-fill and interpolation by lowpass filtering.

Interpolation Filter Design

Let $X(f)$ be the Fourier transform of any function, $x(t)$, whose samples at some interval, T, equal the $x[n]$ sequence. Then the discrete-time Fourier transform (DTFT) of the $x[n]$ sequence is the Fourier series representation of a periodic summation of $X(f)$:

$$\underbrace{\sum_{n=-\infty}^{\infty} \overbrace{x(nT)}^{x[n]}\, e^{-i2\pi fnT}}_{\text{DTFT}} = \frac{1}{T}\sum_{k=-\infty}^{\infty} X(f - k/T).$$

When T has units of seconds, f has units of hertz. Sampling L times faster (at interval T/L) increases the periodicity by a factor of L:

$$\frac{L}{T}\sum_{k=-\infty}^{\infty} X\left(f-k\cdot\frac{L}{T}\right),$$

which is also the desired result of interpolation. An example of both these distributions is depicted in the top two graphs of Fiure.

When the additional samples are inserted zeros, they increase the data rate, but they have no effect on the frequency distribution until the zeros are replaced by the interpolation filter. Many filter design programs use frequency units of *cycles/sample*, which is achieved by normalizing the frequency axis, based on the new data rate (L/T). The result is shown in the third graph of figure. Also shown is the passband of the interpolation filter needed to make the third graph resemble the second one. Its cutoff frequency is $\frac{0.5}{L}$. In terms of actual frequency, the cutoff is $\frac{0.5}{T}$. Hz, which is the Nyquist frequency of the original x[n] sequence.

The same result can be obtained from Z-transforms, constrained to values of complex-variable, z, of the form $z=e^{i\omega}$. Then the transform is the same Fourier series with different frequency normalization. By comparison with Eq.1, we deduce:

$$\sum_{n=-\infty}^{\infty} x[n]\,z^{-n} = \sum_{n=-\infty}^{\infty} x[n]\,e^{-i\omega n} = \frac{1}{T}\sum_{k=-\infty}^{\infty} \underbrace{X\left(\frac{\omega}{2\pi T}-\frac{k}{T}\right)}_{X\left(\frac{\omega-2\pi k}{2\pi T}\right)}$$

which is depicted by the fourth graph in figure. When the zeros are inserted, the transform becomes:

$$\sum_{n=-\infty}^{\infty} x[n]\,z^{-nL} = \sum_{n=-\infty}^{\infty} x[n]\,e^{-i\omega Ln} = \frac{1}{T}\sum_{k=-\infty}^{\infty} \underbrace{X\left(\tfrac{\omega L}{2\pi T}-\tfrac{k}{T}\right)}_{X\left(\frac{\omega-2\pi k/L}{2\pi T/L}\right)},$$

depicted by the bottom graph. In these normalizations, the effective data-rate is always represented by the constant 2π (*radians/sample*) instead of 1. In those units, the interpolation filter bandwidth is π/L, as show on the bottom graph. The corresponding physical frequency is $\frac{}{L}\cdot\frac{}{T}=\frac{0.5}{T}$ Hz, the original Nyquist frequency.

Upsampling by a Rational Fraction

Let L/M denote the upsampling factor, where $L > M$.

- Upsample by a factor of L.
- Downsample by a factor of M.

- Upsampling requires a lowpass filter after increasing the data rate, and downsampling requires a lowpass filter before decimation. Therefore, both operations can be accomplished by a single filter with the lower of the two cutoff frequencies. For the $L > M$ case, the interpolation filter cutoff, $\frac{0.5}{L}$ *cycles per intermediate sample*, is the lower frequency.

Analog Signal Processing

Analog signal processing is a type of signal processing conducted on continuous analog signals by some analog means (as opposed to the discrete digital signal processing where the signal processing is carried out by a digital process). "Analog" indicates something that is mathematically represented as a set of continuous values. This differs from "digital" which uses a series of discrete quantities to represent signal. Analog values are typically represented as a voltage, electric current, or electric charge around components in the electronic devices. An error or noise affecting such physical quantities will result in a corresponding error in the signals represented by such physical quantities.

Examples of *analog signal processing* include crossover filters in loudspeakers, "bass", "treble" and "volume" controls on stereos, and "tint" controls on TVs. Common analog processing elements include capacitors, resistors and inductors (as the passive elements) and transistors or opamps (as the active elements).

Tools used in Analog Signal Processing

A system's behavior can be mathematically modeled and is represented in the time domain as h(t) and in the frequency domain as H(s), where s is a complex number in the form of s=a+ib, or s=a+jb in electrical engineering terms (electrical engineers use "j" instead of "i" because current is represented by the variable i). Input signals are usually called x(t) or X(s) and output signals are usually called y(t) or Y(s).

Convolution

Convolution is the basic concept in signal processing that states an input signal can be combined with the system's function to find the output signal. It is the integral of the product of two waveforms after one has reversed and shifted; the symbol for convolution is *.

$$y(t) = (x * h)(t) = \int_a^b x(\tau)h(t-\tau)d\tau$$

That is the convolution integral and is used to find the convolution of a signal and a system; typically a = -∞ and b = +∞.

Consider two waveforms f and g. By calculating the convolution, we determine how much a reversed function g must be shifted along the x-axis to become identical to function f. The convolution function essentially reverses and slides function g along the axis, and calculates the integral of their (f and the reversed and shifted g) product for each possible amount of sliding. When the functions match, the value of (f*g) is maximized. This occurs because when positive areas (peaks) or negative areas (troughs) are multiplied, they contribute to the integral.

Fourier Transform

The Fourier transform is a function that transforms a signal or system in the time domain into the frequency domain, but it only works for certain functions. The constraint on which systems or signals can be transformed by the Fourier Transform is that:

$$\int_{-\infty}^{\infty} |x(t)|dt < \infty$$

This is the Fourier transform integral:

$$X(j\omega) = \int_{-\infty}^{\infty} x(t)e^{-j\omega t}\, dt$$

Usually the Fourier transform integral isn't used to determine the transform; instead, a table of transform pairs is used to find the Fourier transform of a signal or system. The inverse Fourier transform is used to go from frequency domain to time domain:

$$x(t) = \frac{1}{2\pi}\int_{-\infty}^{\infty} X(j\omega)e^{j\omega t}\, d\omega$$

Each signal or system that can be transformed has a unique Fourier transform. There is only one time signal for any frequency signal, and vice versa.

Laplace Transform

The Laplace transform is a generalized Fourier transform. It allows a transform of any system or signal because it is a transform into the complex plane instead of just the jω line like the Fourier transform. The major difference is that the Laplace transform has a region of convergence for which the transform is valid. This implies that a signal in frequency may have more than one signal in time; the correct time signal for the transform is determined by the region of convergence. If the region of convergence includes the jω axis, jω can be substituted into the Laplace transform for s and it's the same as the Fourier transform. The Laplace transform is:

$$X(s) = \int_{0^-}^{\infty} x(t)e^{-st}\, dt$$

and the inverse Laplace transform, if all the singularities of X(s) are in the left half of the complex plane, is:

$$x(t) = \frac{1}{2\pi}\int_{-\infty}^{\infty} X(s)e^{st}\, ds$$

Bode Plots

Bode plots are plots of magnitude vs. frequency and phase vs. frequency for a system. The magnitude axis is in [Decibel] (dB). The phase axis is in either degrees or radians. The frequency axes are in a [logarithmic scale]. These are useful because for sinusoidal inputs, the output is the input multiplied by the value of the magnitude plot at the frequency and shifted by the value of the phase plot at the frequency.

Domains

Time Domain

This is the domain that most people are familiar with. A plot in the time domain shows the amplitude of the signal with respect to time.

Frequency Domain

A plot in the frequency domain shows either the phase shift or magnitude of a signal at each frequency that it exists at. These can be found by taking the Fourier transform of a time signal and are plotted similarly to a bode plot.

Signals

While any signal can be used in analog signal processing, there are many types of signals that are used very frequently.

Sinusoids

Sinusoids are the building block of analog signal processing. All real world signals can be represented as an infinite sum of sinusoidal functions via a Fourier series. A sinusoidal function can be represented in terms of an exponential by the application of Euler's Formula.

Impulse

An impulse (Dirac delta function) is defined as a signal that has an infinite magnitude and an infinitesimally narrow width with an area under it of one, centered at zero. An impulse can be represented as an infinite sum of sinusoids that includes all possible frequencies. It is not, in reality, possible to generate such a signal, but it can be sufficiently approximated with a large amplitude, narrow pulse, to produce the theoretical impulse response in a network to a high degree of accuracy. The symbol for an impulse is $\delta(t)$. If an impulse is used as an input to a system, the output is known as the impulse response. The impulse response defines the system because all possible frequencies are represented in the input

Step

A unit step function, also called the Heaviside step function, is a signal that has a magnitude of zero before zero and a magnitude of one after zero. The symbol for a unit step is $u(t)$. If a step is used as the input to a system, the output is called the step response. The step response shows how a system responds to a sudden input, similar to turning on a switch. The period before the output stabilizes is called the transient part of a signal. The step response can be multiplied with other signals to show how the system responds when an input is suddenly turned on.

The unit step function is related to the Dirac delta function by:

$$u(t) = \int_{-\infty}^{t} \delta(s)ds$$

Systems

Linear Time-invariant (LTI)

Linearity means that if you have two inputs and two corresponding outputs, if you take a linear combination of those two inputs you will get a linear combination of the outputs. An example of a linear system is a first order low-pass or high-pass filter. Linear systems are made out of analog devices that demonstrate linear properties. These devices don't have to be entirely linear, but must have a region of operation that is linear. An operational amplifier is a non-linear device, but has a region of operation that is linear, so it can be modeled as linear within that region of operation. Time-invariance means it doesn't matter when you start a system, the same output will result. For example, if you have a system and put an input into it today, you would get the same output if you started the system tomorrow instead. There aren't any real systems that are LTI, but many systems can be modeled as LTI for simplicity in determining what their output will be. All systems have some dependence on things like temperature, signal level or other factors that cause them to be non-linear or non-time-invariant, but most are stable enough to model as LTI. Linearity and time-invariance are important because they are the only types of systems that can be easily solved using conventional analog signal processing methods. Once a system becomes non-linear or non-time-invariant, it becomes a non-linear differential equations problem, and there are very few of those that can actually be solved.

Signal Sampling

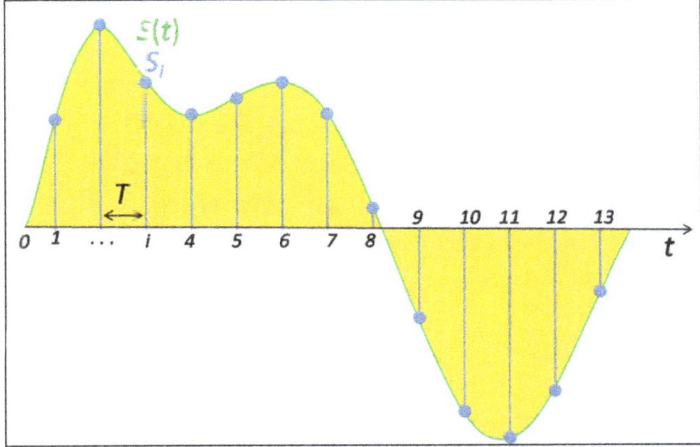

Signal sampling representation. The continuous signal is represented with a green colored line while the discrete samples are indicated by the blue vertical lines.

In signal processing, sampling is the reduction of a continuous-time signal to a discrete-time signal. A common example is the conversion of a sound wave (a continuous signal) to a sequence of samples (a discrete-time signal).

A sample is a value or set of values at a point in time and/or space.

A sampler is a subsystem or operation that extracts samples from a continuous signal.

A theoretical ideal sampler produces samples equivalent to the instantaneous value of the continuous signal at the desired points.

The original signal is retrievable from a sequence of samples, up to the Nyquist limit, by passing the sequence of samples through a type of low pass filter called a reconstruction filter.

Theory

Sampling can be done for functions varying in space, time, or any other dimension, and similar results are obtained in two or more dimensions.

For functions that vary with time, let $s(t)$ be a continuous function (or "signal") to be sampled, and let sampling be performed by measuring the value of the continuous function every T seconds, which is called the sampling interval or the sampling period. Then the sampled function is given by the sequence:

$s(nT)$, for integer values of n.

The sampling frequency or sampling rate, f_s, is the average number of samples obtained in one second (*samples per second*), thus $f_s = 1/T$.

Reconstructing a continuous function from samples is done by interpolation algorithms. The Whittaker–Shannon interpolation formula is mathematically equivalent to an ideal lowpass filter whose input is a sequence of Dirac delta functions that are modulated (multiplied) by the sample values. When the time interval between adjacent samples is a constant (T), the sequence of delta functions is called a Dirac comb. Mathematically, the modulated Dirac comb is equivalent to the product of the comb function with $s(t)$. That purely mathematical abstraction is sometimes referred to as *impulse sampling*.

Most sampled signals are not simply stored and reconstructed. But the fidelity of a theoretical reconstruction is a customary measure of the effectiveness of sampling. That fidelity is reduced when $s(t)$ contains frequency components whose periodicity is smaller than two samples; or equivalently the ratio of cycles to samples exceeds ½ . The quantity ½ *cycles/sample* × f_s *samples/sec* = $f_s/2$ *cycles/sec* (hertz) is known as the Nyquist frequency of the sampler. Therefore, $s(t)$ is usually the output of a lowpass filter, functionally known as an *anti-aliasing filter*. Without an anti-aliasing filter, frequencies higher than the Nyquist frequency will influence the samples in a way that is misinterpreted by the interpolation process.

Practical Considerations

In practice, the continuous signal is sampled using an analog-to-digital converter (ADC), a device with various physical limitations. This results in deviations from the theoretically perfect reconstruction, collectively referred to as distortion.

Various types of distortion can occur, including:

- Aliasing: Some amount of aliasing is inevitable because only theoretical, infinitely long, functions can have no frequency content above the Nyquist frequency. Aliasing can be made arbitrarily small by using a sufficiently large order of the anti-aliasing filter.

- Aperture error results from the fact that the sample is obtained as a time average within a sampling region, rather than just being equal to the signal value at the sampling instant. In a capacitor-based sample and hold circuit, aperture errors are introduced by multiple mechanisms. For example, the capacitor cannot instantly track the input signal and the capacitor can not instantly be isolated from the input signal.
- Jitter or deviation from the precise sample timing intervals.
- Noise, including thermal sensor noise, analog circuit noise, etc.
- Slew rate limit error, caused by the inability of the ADC input value to change sufficiently rapidly.
- Quantization as a consequence of the finite precision of words that represent the converted values.
- Error due to other non-linear effects of the mapping of input voltage to converted output value (in addition to the effects of quantization).

Although the use of oversampling can completely eliminate aperture error and aliasing by shifting them out of the pass band, this technique cannot be practically used above a few GHz, and may be prohibitively expensive at much lower frequencies. Furthermore, while oversampling can reduce quantization error and non-linearity, it cannot eliminate these entirely. Consequently, practical ADCs at audio frequencies typically do not exhibit aliasing, aperture error, and are not limited by quantization error. Instead, analog noise dominates. At RF and microwave frequencies where oversampling is impractical and filters are expensive, aperture error, quantization error and aliasing can be significant limitations.

Jitter, noise, and quantization are often analyzed by modeling them as random errors added to the sample values. Integration and zero-order hold effects can be analyzed as a form of low-pass filtering. The non-linearities of either ADC or DAC are analyzed by replacing the ideal linear function mapping with a proposed nonlinear function.

Applications

Audio Sampling

Digital audio uses pulse-code modulation and digital signals for sound reproduction. This includes analog-to-digital conversion (ADC), digital-to-analog conversion (DAC), storage, and transmission. In effect, the system commonly referred to as digital is in fact a discrete-time, discrete-level analog of a previous electrical analog. While modern systems can be quite subtle in their methods, the primary usefulness of a digital system is the ability to store, retrieve and transmit signals without any loss of quality.

Sampling Rate

A commonly seen unit of sampling rate is Hz, which stands for Hertz and means "samples per second". As an example, 48 kHz is 48,000 samples per second.

When it is necessary to capture audio covering the entire 20–20,000 Hz range of human hearing, such as when recording music or many types of acoustic events, audio waveforms are typically sampled at 44.1 kHz (CD), 48 kHz, 88.2 kHz, or 96 kHz. The approximately double-rate requirement is a consequence of the Nyquist theorem. Sampling rates higher than about 50 kHz to 60 kHz cannot supply more usable information for human listeners. Early professional audio equipment manufacturers chose sampling rates in the region of 40 to 50 kHz for this reason.

There has been an industry trend towards sampling rates well beyond the basic requirements: such as 96 kHz and even 192 kHz Even though ultrasonic frequencies are inaudible to humans, recording and mixing at higher sampling rates is effective in eliminating the distortion that can be caused by foldback aliasing. Conversely, ultrasonic sounds may interact with and modulate the audible part of the frequency spectrum (intermodulation distortion), *degrading* the fidelity. One advantage of higher sampling rates is that they can relax the low-pass filter design requirements for ADCs and DACs, but with modern oversampling sigma-delta converters this advantage is less important.

The Audio Engineering Society recommends 48 kHz sampling rate for most applications but gives recognition to 44.1 kHz for Compact Disc (CD) and other consumer uses, 32 kHz for transmission-related applications, and 96 kHz for higher bandwidth or relaxed anti-aliasing filtering. Both Lavry Engineering and J. Robert Stuart state that the ideal sampling rate would be about 60 kHz, but since this is not a standard frequency, recommend 88.2 or 96 kHz for recording purposes.

A more complete list of common audio sample rates is:

Sampling rate	Use
8,000 Hz	Telephone and encrypted walkie-talkie, wireless intercom and wireless microphone transmission; adequate for human speech but without sibilance (ess sounds like eff (/s/, /f/)).
11,025 Hz	One quarter the sampling rate of audio CDs; used for lower-quality PCM, MPEG audio and for audio analysis of subwoofer bandpasses.
16,000 Hz	Wideband frequency extension over standard telephone narrowband 8,000 Hz. Used in most modern VoIP and VVoIP communication products.
22,050 Hz	One half the sampling rate of audio CDs; used for lower-quality PCM and MPEG audio and for audio analysis of low frequency energy. Suitable for digitizing early 20th century audio formats such as 78s.
32,000 Hz	miniDV digital video camcorder, video tapes with extra channels of audio (e.g. DVCAM with four channels of audio), DAT (LP mode), Germany's Digitales Satellitenradio, NICAM digital audio, used alongside analogue television sound in some countries. High-quality digital wireless microphones. Suitable for digitizing FM radio.
37,800 Hz	CD-XA audio.
44,056 Hz	Used by digital audio locked to NTSC color video signals (3 samples per line, 245 lines per field, 59.94 fields per second = 29.97 frames per second).
44,100 Hz	Audio CD, also most commonly used with MPEG-1 audio (VCD, SVCD, MP3). Originally chosen by Sony because it could be recorded on modified video equipment running at either 25 frames per second (PAL) or 30 frame/s (using an NTSC monochrome video recorder) and cover the 20 kHz bandwidth thought necessary to match professional analog recording equipment of the time. A PCM adaptor would fit digital audio samples into the analog video channel of, for example, PAL video tapes using 3 samples per line, 588 lines per frame, 25 frames per second.
47,250 Hz	world's first commercial PCM sound recorder by Nippon Columbia (Denon).

48,000 Hz	The standard audio sampling rate used by professional digital video equipment such as tape recorders, video servers, vision mixers and so on. This rate was chosen because it could reconstruct frequencies up to 22 kHz and work with 29.97 frames per second NTSC video – as well as 25 frame/s, 30 frame/s and 24 frame/s systems. With 29.97 frame/s systems it is necessary to handle 1601.6 audio samples per frame delivering an integer number of audio samples only every fifth video frame. Also used for sound with consumer video formats like DV, digital TV, DVD, and films. The professional Serial Digital Interface (SDI) and High-definition Serial Digital Interface (HD-SDI) used to connect broadcast television equipment together uses this audio sampling frequency. Most professional audio gear uses 48 kHz sampling, including mixing consoles, and digital recording devices.
50,000 Hz	First commercial digital audio recorders from the late 70s from 3M and Soundstream.
50,400 Hz	Sampling rate used by the Mitsubishi X-80 digital audio recorder.
64,000 Hz	Uncommonly used, but supported by some hardware and software.
88,200 Hz	Sampling rate used by some professional recording equipment when the destination is CD (multiples of 44,100 Hz). Some pro audio gear uses (or is able to select) 88.2 kHz sampling, including mixers, EQs, compressors, reverb, crossovers and recording devices.
96,000 Hz	DVD-Audio, some LPCM DVD tracks, BD-ROM (Blu-ray Disc) audio tracks, HD DVD (High-Definition DVD) audio tracks. Some professional recording and production equipment is able to select 96 kHz sampling. This sampling frequency is twice the 48 kHz standard commonly used with audio on professional equipment.
176,400 Hz	Sampling rate used by HDCD recorders and other professional applications for CD production. Four times the frequency of 44.1 kHz.
192,000 Hz	DVD-Audio, some LPCM DVD tracks, BD-ROM (Blu-ray Disc) audio tracks, and HD DVD (High-Definition DVD) audio tracks, High-Definition audio recording devices and audio editing software. This sampling frequency is four times the 48 kHz standard commonly used with audio on professional video equipment.
352,800 Hz	Digital eXtreme Definition, used for recording and editing Super Audio CDs, as 1-bit Direct Stream Digital (DSD) is not suited for editing. Eight times the frequency of 44.1 kHz.
2,822,400 Hz	SACD, 1-bit delta-sigma modulation process known as Direct Stream Digital, co-developed by Sony and Philips.
5,644,800 Hz	Double-Rate DSD, 1-bit Direct Stream Digital at 2× the rate of the SACD. Used in some professional DSD recorders.
11,289,600 Hz	Quad-Rate DSD, 1-bit Direct Stream Digital at 4× the rate of the SACD. Used in some uncommon professional DSD recorders.
22,579,200 Hz	Octuple-Rate DSD, 1-bit Direct Stream Digital at 8× the rate of the SACD. Used in rare experimental DSD recorders. Also known as DSD512.

Bit Depth

Audio is typically recorded at 8-, 16-, and 24-bit depth, which yield a theoretical maximum signal-to-quantization-noise ratio (SQNR) for a pure sine wave of, approximately, 49.93 dB, 98.09 dB and 122.17 dB. CD quality audio uses 16-bit samples. Thermal noise limits the true number of bits that can be used in quantization. Few analog systems have signal to noise ratios (SNR) exceeding 120 dB. However, digital signal processing operations can have very high dynamic range, consequently it is common to perform mixing and mastering operations at 32-bit precision and then convert to 16- or 24-bit for distribution.

Speech Sampling

Speech signals, i.e., signals intended to carry only human speech, can usually be sampled at a much lower rate. For most phonemes, almost all of the energy is contained in the 100 Hz–4 kHz range, allowing a sampling rate of 8 kHz. This is the sampling rate used by nearly all telephony systems, which use the G.711 sampling and quantization specifications.

Video Sampling

Standard-definition television (SDTV) uses either 720 by 480 pixels (US NTSC 525-line) or 720 by 576 pixels (UK PAL 625-line) for the visible picture area.

High-definition television (HDTV) uses 720p (progressive), 1080i (interlaced), and 1080p (progressive, also known as Full-HD).

In digital video, the temporal sampling rate is defined the frame rate – or rather the field rate – rather than the notional pixel clock. The image sampling frequency is the repetition rate of the sensor integration period. Since the integration period may be significantly shorter than the time between repetitions, the sampling frequency can be different from the inverse of the sample time:

- 50 Hz – PAL video.
- 60 / 1.001 Hz ~= 59.94 Hz – NTSC video.

Video digital-to-analog converters operate in the megahertz range (from ~3 MHz for low quality composite video scalers in early games consoles, to 250 MHz or more for the highest-resolution VGA output).

When analog video is converted to digital video, a different sampling process occurs, this time at the pixel frequency, corresponding to a spatial sampling rate along scan lines. A common pixel sampling rate is:

- 13.5 MHz – CCIR 601, D1 video.

Spatial sampling in the other direction is determined by the spacing of scan lines in the raster. The sampling rates and resolutions in both spatial directions can be measured in units of lines per picture height.

Spatial aliasing of high-frequency luma or chroma video components shows up as a moiré pattern.

3D Sampling

The process of volume rendering samples a 3D grid of voxels to produce 3D renderings of sliced (tomographic) data. The 3D grid is assumed to represent a continuous region of 3D space. Volume rendering is common in medial imaging, X-ray computed tomography (CT/CAT), magnetic resonance imaging (MRI), positron emission tomography (PET) are some examples. It is also used for seismic tomography and other applications.

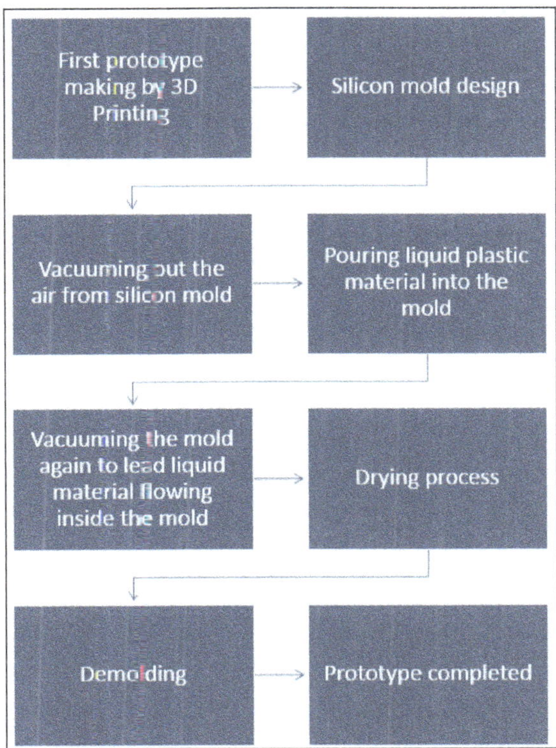

Complex Sampling

Complex sampling (I/Q sampling) is the simultaneous sampling of two different, but related, waveforms, resulting in pairs of samples that are subsequently treated as complex numbers. When one waveform, $\hat{s}(t)$, is the Hilbert transform of the other waveform, $s(t)$, the complex-valued function, $s_a(t) \triangleq s(t) + i \cdot \hat{s}(t)$, is called an analytic signal, whose Fourier transform is zero for all negative values of frequency. In that case, the Nyquist rate for a waveform with no frequencies ≥ *B* can be reduced to just *B* (complex samples/sec), instead of 2*B* (real samples/sec). More apparently, the equivalent baseband waveform, $s_a(t) \cdot e^{-i2\pi \frac{B}{2} t}$, also has a Nyquist rate of *B*, because all of its non-zero frequency content is shifted into the interval [-B/2, B/2).

Although complex-valued samples can be obtained as described above, they are also created by manipulating samples of a real-valued waveform. For instance, the equivalent baseband waveform can be created without explicitly computing $\hat{s}(t)$, by processing the product sequence, $\left[s(nT) \cdot e^{-i2\pi \frac{B}{2} Tn} \right]$, through a digital lowpass filter whose cutoff frequency is B/2. Computing only every other sample of the output sequence reduces the sample-rate commensurate with the reduced Nyquist rate. The result is half as many complex-valued samples as the original number of real samples. No information is lost, and the original s(t) waveform can be recovered, if necessary.

Undersampling

In signal processing, undersampling or bandpass sampling is a technique where one samples a bandpass-filtered signal at a sample rate below its Nyquist rate (twice the upper cutoff frequency), but is still able to reconstruct the signal.

When one undersamples a bandpass signal, the samples are indistinguishable from the samples of a low-frequency alias of the high-frequency signal. Such sampling is also known as bandpass sampling, harmonic sampling, IF sampling, and direct IF-to-digital conversion.

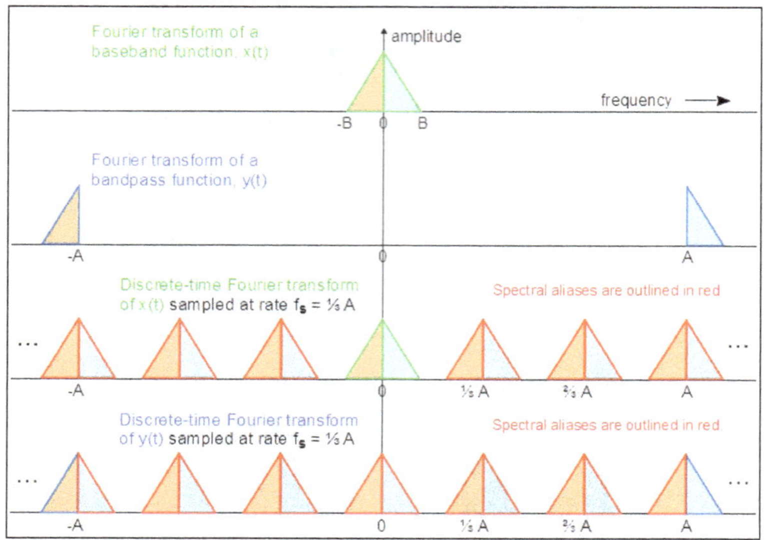

The top 2 graphs depict Fourier transforms of 2 different functions that produce the same results when sampled at a particular rate. The baseband function is sampled faster than its Nyquist rate, and the bandpass function is undersampled, effectively converting it to baseband. The lower graphs indicate how identical spectral results are created by the aliases of the sampling process.

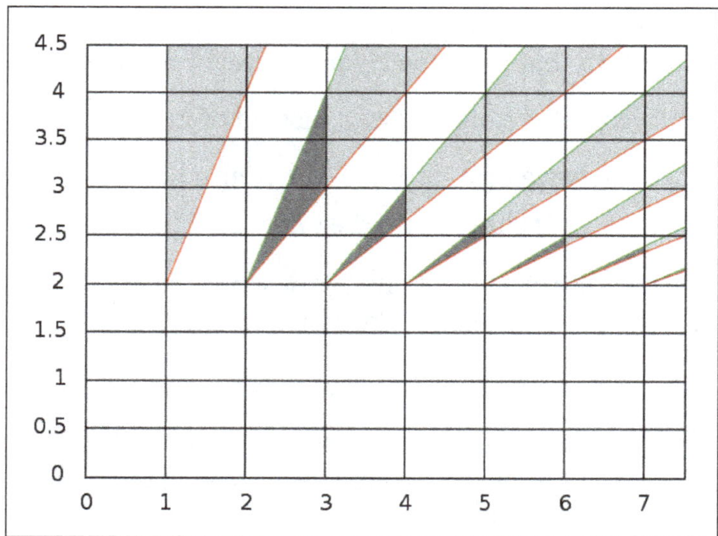

Plot of sample rates (y axis) versus the upper edge frequency (x axis) for a band of width 1; grays areas are combinations that are "allowed" in the sense that no two frequencies in the band alias to same frequency. The darker gray areas correspond to undersampling with the maximum value of n in the equations.

The Fourier transforms of real-valued functions are symmetrical around the 0 Hz axis. After sampling, only a periodic summation of the Fourier transform (called discrete-time Fourier transform) is still available. The individual frequency-shifted copies of the original transform are called *aliases*. The frequency offset between adjacent aliases is the sampling-rate, denoted by f_s. When the aliases are mutually exclusive (spectrally), the original transform and the original continuous function, or a frequency-shifted version of it (if desired), can be recovered from the samples. The first and third graphs of figure depict a baseband spectrum before and after being sampled at a rate that completely separates the aliases.

The second graph of figure depicts the frequency profile of a bandpass function occupying the band (A, $A+B$) (shaded blue) and its mirror image (shaded beige). The condition for a non-destructive sample rate is that the aliases of both bands do not overlap when shifted by all integer multiples of f_s. The fourth graph depicts the spectral result of sampling at the same rate as the baseband function. The rate was chosen by finding the lowest rate that is an integer sub-multiple of A and also satisfies the baseband Nyquist criterion: $f_s > 2B$. Consequently, the bandpass function has effectively been converted to baseband. All the other rates that avoid overlap are given by these more general criteria, where A and $A+B$ are replaced by f_L and f_H, respectively:

$$\frac{2f_H}{n} \le f_s \le \frac{2f_L}{n-1}, \text{ for any integer } n \text{ satisfying: } 1 \le n \le \left\lfloor \frac{f_H}{f_H - f_L} \right\rfloor.$$

The highest n for which the condition is satisfied leads to the lowest possible sampling rates.

Important signals of this sort include a radio's intermediate-frequency (IF), radio-frequency (RF) signal, and the individual channels of a filter bank.

If $n > 1$, then the conditions result in what is sometimes referred to as undersampling, bandpass sampling, or using a sampling rate less than the Nyquist rate ($2f_H$). For the case of a given sampling frequency, simpler formulae for the constraints on the signal's spectral band are given below.

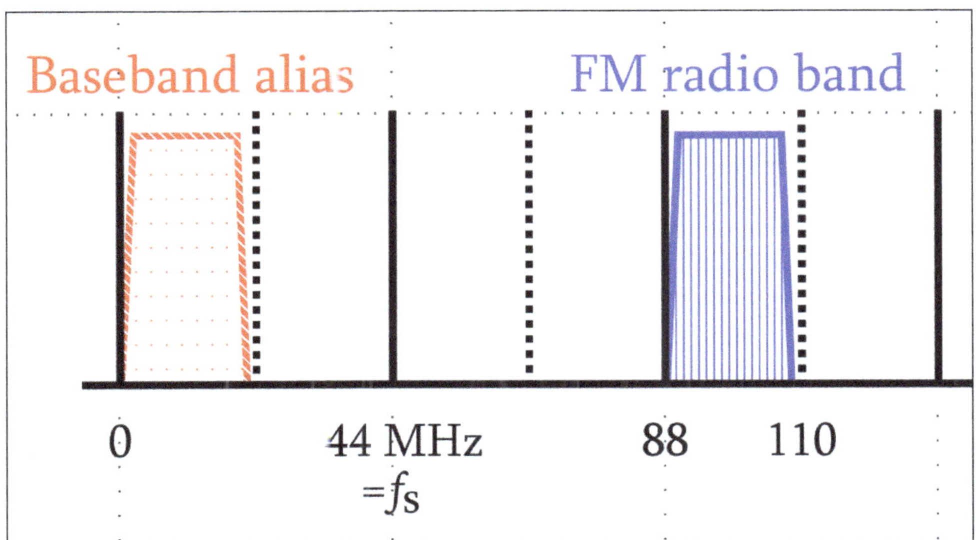

Spectrum of the FM radio band (88–108 MHz) and its baseband alias under 44 MHz ($n = 5$) sampling. An anti-alias filter quite tight to the FM radio band is required, and there's not room for stations at nearby expansion channels such as 87.9 without aliasing.

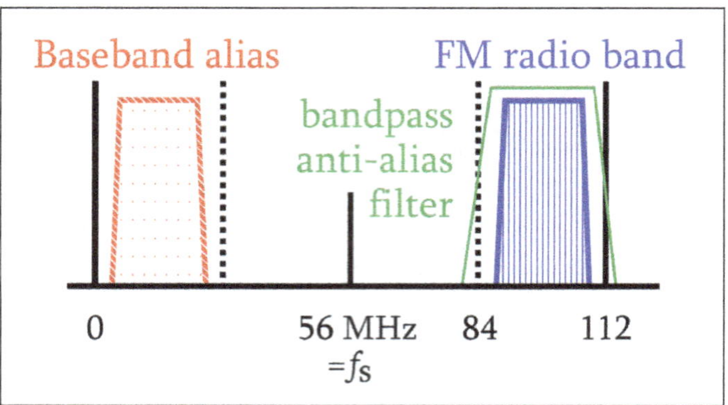

Spectrum of the FM radio band (88–108 MHz) and its baseband alias under 56 MHz (n = 4) sampling, showing plenty of room for bandpass anti-aliasing filter transition bands. The baseband image is frequency-reversed in this case (even n).

Example: Consider FM radio to illustrate the idea of undersampling.

In the US, FM radio operates on the frequency band from f_L = 88 MHz to f_H = 108 MHz. The bandwidth is given by

$$W = f_H - f_L = 108\text{ MHz} - 88\text{ MHz} = 20\text{ MHz}$$

The sampling conditions are satisfied for:

$$1 \le n \le \lfloor 5.4 \rfloor = \left\lfloor \frac{108\text{ MHz}}{20\text{ MHz}} \right\rfloor$$

Therefore, n can be 1, 2, 3, 4, or 5.

The value n = 5 gives the lowest sampling frequencies interval $43.2\text{ MHz} < f_s < 44\text{ MHz}$ and this is a scenario of undersampling. In this case, the signal spectrum fits between 2 and 2.5 times the sampling rate (higher than 86.4–88 MHz but lower than 108–110 MHz).

A lower value of n will also lead to a useful sampling rate. For example, using n = 4, the FM band spectrum fits easily between 1.5 and 2.0 times the sampling rate, for a sampling rate near 56 MHz (multiples of the Nyquist frequency being 28, 56, 84, 112, etc.).

When undersampling a real-world signal, the sampling circuit must be fast enough to capture the highest signal frequency of interest. Theoretically, each sample should be taken during an infinitesimally short interval, but this is not practically feasible. Instead, the sampling of the signal should be made in a short enough interval that it can represent the instantaneous value of the signal with the highest frequency. This means that in the FM radio example above, the sampling circuit must be able to capture a signal with a frequency of 108 MHz, not 43.2 MHz. Thus, the sampling frequency may be only a little bit greater than 43.2 MHz, but the input bandwidth of the system must be at least 108 MHz. Similarly, the accuracy of the sampling timing, or aperture uncertainty of the sampler, frequently the analog-to-digital converter, must be appropriate for the frequencies being sampled 108MHz, not the lower sample rate.

If the sampling theorem is interpreted as requiring twice the highest frequency, then the required sampling rate would be assumed to be greater than the Nyquist rate 216 MHz. While this does satisfy the last condition on the sampling rate, it is grossly oversampled.

Note that if a band is sampled with $n > 1$, then a band-pass filter is required for the anti-aliasing filter, instead of a lowpass filter.

As we have seen, the normal baseband condition for reversible sampling is that $X(f) = 0$ outside the interval: $\left(-\frac{1}{2}f_s, \frac{1}{2}f_s\right)$,
and the reconstructive interpolation function, or lowpass filter impulse response, is $sinc(t/T)$.

To accommodate undersampling, the bandpass condition is that $X(f) = 0$ outside the union of open positive and negative frequency bands:

$\left(-\frac{n}{2}f_s, -\frac{n-1}{2}f_s\right) \cup \left(\frac{n-1}{2}f_s, \frac{n}{2}f_s\right)$ for some positive integer n.

which includes the normal baseband condition as case $n = 1$ (except that where the intervals come together at 0 frequency, they can be closed).

The corresponding interpolation function is the bandpass filter given by this difference of lowpass impulse responses:

$$n\,\mathrm{sinc}\left(\frac{nt}{T}\right) - (n-1)\,\mathrm{sinc}\left(\frac{(n-1)t}{T}\right).$$

On the other hand, reconstruction is not usually the goal with sampled IF or RF signals. Rather, the sample sequence can be treated as ordinary samples of the signal frequency-shifted to near baseband, and digital demodulation can proceed on that basis, recognizing the spectrum mirroring when n is even.

Further generalizations of undersampling for the case of signals with multiple bands are possible, and signals over multidimensional domains (space or space-time) and have been worked out in detail by Igor Kluvánek.

Oversampling

In signal processing, oversampling is the process of sampling a signal at a sampling frequency significantly higher than the Nyquist rate. Theoretically, a bandwidth-limited signal can be perfectly reconstructed if sampled at the Nyquist rate or above it. The Nyquist rate is defined as twice the highest frequency component in the signal. Oversampling is capable of improving resolution and signal-to-noise ratio, and can be helpful in avoiding aliasing and phase distortion by relaxing anti-aliasing filter performance requirements.

A signal is said to be oversampled by a factor of N if it is sampled at N times the Nyquist rate.

Motivation

There are three main reasons for performing oversampling,

Anti-aliasing

Oversampling can make it easier to realize analog anti-aliasing filters. Without oversampling, it is very difficult to implement filters with the sharp cutoff necessary to maximize use of the available bandwidth without exceeding the Nyquist limit. By increasing the bandwidth of the sampling system, design constraints for the anti-aliasing filter may be relaxed. Once sampled, the signal can be digitally filtered and downsampled to the desired sampling frequency. In modern integrated circuit technology, the digital filter associated with this downsampling are easier to implement than a comparable analog filter required by a non-oversampled system.

Resolution

In practice, oversampling is implemented in order to reduce cost and improve performance of an analog-to-digital converter (ADC) or digital-to-analog converter (DAC). When oversampling by a factor of N, the dynamic range also increases a factor of N because there are N times as many possible values for the sum. However, the signal-to-noise ratio (SNR) increases by $\sqrt{N}$, because summing up uncorrelated noise increases its amplitude by $\sqrt{N}$, while summing up a coherent signal increases its average by N. As a result, the SNR increases by $\sqrt{N}$.

For instance, to implement a 24-bit converter, it is sufficient to use a 20-bit converter that can run at 256 times the target sampling rate. Combining 256 consecutive 20-bit samples can increase the SNR by a factor of 16, effectively adding 4 bits to the resolution and producing a single sample with 24-bit resolution. While with N=256 there is an increase in dynamic range by 8 bits, and the level of coherent signal increases by a factor of N, the noise changes by a factor of $\sqrt{N}$ =16, so the net SNR improves by a factor of 16, 4 bits or 24 dB.

The number of samples required to get n bits of additional data precision is:

$$\text{number of samples} = (2^n)^2 = 2^{2n}.$$

To get the mean sample scaled up to an integer with n additional bits, the sum of 2^{2n} samples is divided by 2^n:

$$\text{scaled mean} = \frac{\sum_{i=0}^{2^{2n}-1} 2^n \text{data}_i}{2^{2n}} = \frac{\sum_{i=0}^{2^{2n}-1} \text{data}_i}{2^n}.$$

This averaging is only effective if the signal contains sufficient uncorrelated noise to be recorded by the ADC. If not, in the case of a stationary input signal, all 2^n samples would have the same value and the resulting average would be identical to this value; so in this case, oversampling would have made no improvement. In similar cases where the ADC records no noise and the input signal is changing over time, oversampling improves the result, but to an inconsistent and unpredictable extent.

Adding some dithering noise to the input signal can actually improve the final result because the dither noise allows oversampling to work to improve resolution. In many practical applications, a small increase in noise is well worth a substantial increase in measurement resolution. In practice, the dithering noise can often be placed outside the frequency range of interest to the measurement, so that this noise can be subsequently filtered out in the digital domain—resulting in a final measurement, in the frequency range of interest, with both higher resolution and lower noise.

Noise

If multiple samples are taken of the same quantity with uncorrelated noise added to each sample, then because, uncorrelated signals combine more weakly than correlated ones, averaging N samples reduces the noise power by a factor of N. If, for example, we oversample by a factor of 4, the signal-to-noise ratio in terms of power improves by factor of 4 which corresponds to a factor of 2 improvement in terms of voltage.

Certain kinds of ADCs known as delta-sigma converters produce disproportionately more quantization noise at higher frequencies. By running these converters at some multiple of the target sampling rate, and low-pass filtering the oversampled signal down to half the target sampling rate, a final result with *less* noise (over the entire band of the converter) can be obtained. Delta-sigma converters use a technique called noise shaping to move the quantization noise to the higher frequencies.

Example:

Consider a signal with a bandwidth or highest frequency of B = 100 Hz. The sampling theorem states that sampling frequency would have to be greater than 200 Hz. Sampling at four times that rate requires a sampling frequency of 800 Hz. This gives the anti-aliasing filter a transition band of 300 Hz ($(f_s/2) - B$ = (800 Hz/2) – 100 Hz = 300 Hz) instead of 0 Hz if the sampling frequency was 200 Hz. Achieving an anti-aliasing filter with 0 Hz transition band is unrealistic whereas an anti-aliasing filter with a transition band of 300 Hz is not difficult.

Oversampling in Reconstruction

The term oversampling is also used to denote a process used in the reconstruction phase of digital-to-analog conversion, in which an intermediate high sampling rate is used between the digital input and the analogue output. Here, samples are interpolated in the digital domain to add additional samples in between, thereby converting the data to a higher sample rate, which is a form of upsampling. When the resulting higher-rate samples are converted to analog, a less complex/ expensive analog low pass filter is required to remove the high-frequency content, which will consist of reflected images of the real signal created by the zero-order hold of the digital-to-analog converter. Essentially, this is a way to shift some of the complexity of the filtering into the digital domain and achieves the same benefit as oversampling in analog-to-digital conversion.

Sampling Rate

Sampling rate determines the sound frequency range (corresponding to pitch) which can be represented in the digital waveform. The range of frequencies represented in a waveform is often called its bandwidth. Waveforms sampled at a high sampling rate can represent a broad range of frequencies and hence have broad bandwidth. In fact, the maximum bandwidth of a sampled waveform is determined exactly by its sampling rate; the maximum frequency representable in a sampled waveform is termed its Nyquist frequency, and is equal to one half the sampling rate. Thus, for example, a waveform sampled at 16,000 Hz can represent all frequencies up to its Nyquist frequency of 8,000 Hz.

A problem called aliasing occurs when a signal to be sampled contains energy at frequencies above the sampling Nyquist frequency. The next figure illustrates how aliasing would occur when the sampling rate is much too low for the frequency of an input signal. The solid curve represents the analog signal at a comparatively high frequency. Circles show where samples were taken at a relatively low sampling rate. The dotted line illustrates the apparent frequency of the sampled waveform, completing about two cycles in the period that the original signal completed 20 cycles.

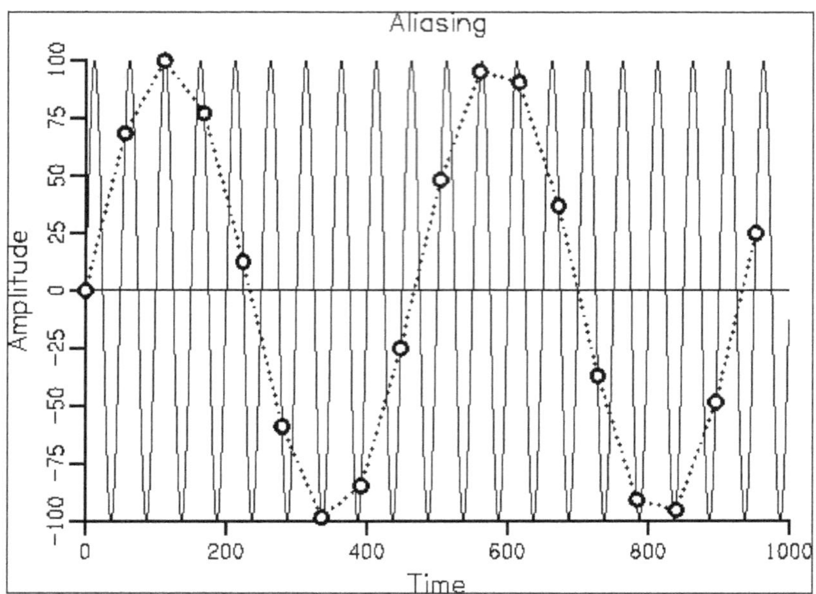

Obviously, aliasing has the effect of producing sounds of lower frequency from sounds that are higher in frequency than the Nyquist frequency. Once aliasing has occurred, it is absolutely impossible to distinguish a component generated by aliasing from one that was actually present in the input signal. This effect is one of the most common sources of distortion in digitized waveforms. Fortunately, most modern computer hardware for digitizing sound has builtin filters which are tuned to remove sound energy at frequencies beyond the Nyquist frequency for whatever sampling rate is being used.

Nyquist–Shannon Sampling Theorem

In the field of digital signal processing, the sampling theorem is a fundamental bridge between continuous-time signals and discrete-time signals. It establishes a sufficient condition for a sample rate that permits a discrete sequence of *samples* to capture all the information from a continuous-time signal of finite bandwidth.

Strictly speaking, the theorem only applies to a class of mathematical functions having a Fourier transform that is zero outside of a finite region of frequencies. Intuitively we expect that when one reduces a continuous function to a discrete sequence and interpolates back to a continuous function, the fidelity of the result depends on the density (or sample rate) of the original samples. The sampling theorem introduces the concept of a sample rate that is sufficient for perfect fidelity for the class of functions that are bandlimited to a given bandwidth, such that no actual information is lost in the sampling process. It expresses the sufficient sample rate in terms of the bandwidth for the class of functions. The theorem also leads to a formula for perfectly reconstructing the original continuous-time function from the samples.

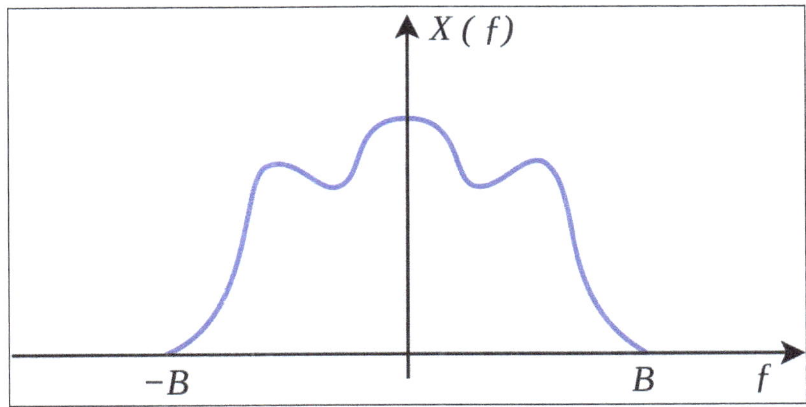

Example of magnitude of the Fourier transform of a bandlimited function.

Perfect reconstruction may still be possible when the sample-rate criterion is not satisfied, provided other constraints on the signal are known. In some cases (when the sample-rate criterion is not satisfied), utilizing additional constraints allows for approximate reconstructions. The fidelity of these reconstructions can be verified and quantified utilizing Bochner's theorem.

The name Nyquist–Shannon sampling theorem honours Harry Nyquist and Claude Shannon albeit the fact that it had already been discovered in 1933 by Vladimir Kotelnikov. The theorem was also discovered independently by E. T. Whittaker and by others. It is thus also known by the names Nyquist–Shannon–Kotelnikov, Whittaker–Shannon–Kotelnikov, Whittaker–Nyquist–Kotelnikov–Shannon, and cardinal theorem of interpolation.

Sampling is a process of converting a signal (for example, a function of continuous time and/or space) into a sequence of values (a function of discrete time and/or space). Shannon's version of the theorem states:

> If a function $x(t)$ contains no frequencies higher than B hertz, it is completely determined by giving its ordinates at a series of points spaced $1/(2B)$ seconds apart.

A sufficient sample-rate is therefore anything larger than $2B$ samples per second. Equivalently, for a given sample rate f_s, perfect reconstruction is guaranteed possible for a bandlimit $B < f_s/2$.

When the bandlimit is too high (or there is no bandlimit), the reconstruction exhibits imperfections known as aliasing. Modern statements of the theorem are sometimes careful to explicitly state that $x(t)$ must contain no sinusoidal component at exactly frequency B, or that B must be strictly less than ½ the sample rate. The threshold $2B$ is called the Nyquist rate and is an attribute of the continuous-time input $x(t)$ to be sampled. The sample rate must exceed the Nyquist rate for the samples to suffice to represent $x(t)$. The threshold $f_s/2$ is called the Nyquist frequency and is an attribute of the sampling equipment. All meaningful frequency components of the properly sampled $x(t)$ exist below the Nyquist frequency. The condition described by these inequalities is called the Nyquist criterion, or sometimes the *Raabe condition*. The theorem is also applicable to functions of other domains, such as *space,* in the case of a digitized image. The only change, in the case of other domains, is the units of measure applied to t, f_s, and B.

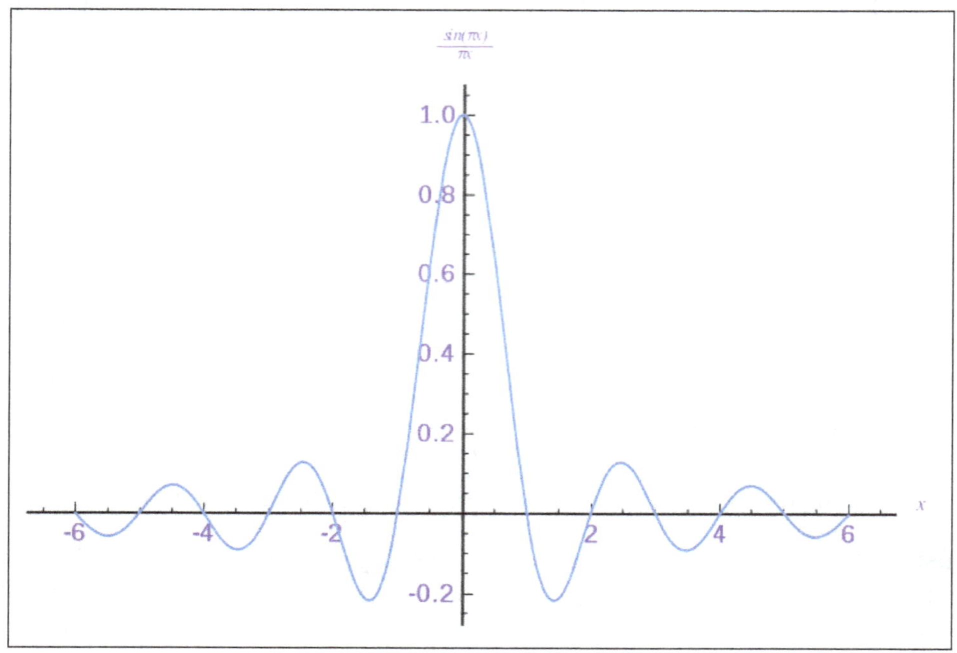

The normalized sinc function: $\sin(\pi x) / (\pi x)$ showing the central peak at $x = 0$, and zero-crossings at the other integer values of x.

The symbol $T = 1/f_s$ is customarily used to represent the interval between samples and is called the sample period or sampling interval. And the samples of function $x(t)$ are commonly denoted by $x[n] = x(nT)$ (alternatively "x_n" in older signal processing literature), for all integer values of n. A mathematically ideal way to interpolate the sequence involves the use of sinc functions. Each sample in the sequence is replaced by a sinc function, centered on the time axis at the original location of the sample, nT, with the amplitude of the sinc function scaled to the sample value, $x[n]$. Subsequently, the sinc functions are summed into a continuous function. A mathematically equivalent method is to convolve one sinc function with a series of Dirac delta pulses, weighted by the sample values. Neither method is numerically practical. Instead, some type of approximation of the sinc functions, finite in length, is used. The imperfections attributable to the approximation are known as *interpolation error.*

Practical digital-to-analog converters produce neither scaled and delayed sinc functions, nor ideal Dirac pulses. Instead they produce a piecewise-constant sequence of scaled and delayed rectangular pulses (the zero-order hold), usually followed by a lowpass filter (called an *"anti-imaging filter"*) to remove spurious high-frequency replicas of the original baseband signal.

When $x(t)$ is a function with a Fourier transform $X(f)$:

$$X(f) \triangleq \int_{-\infty}^{\infty} x(t)\, e^{-i2\pi ft}\, \mathrm{d}t,$$

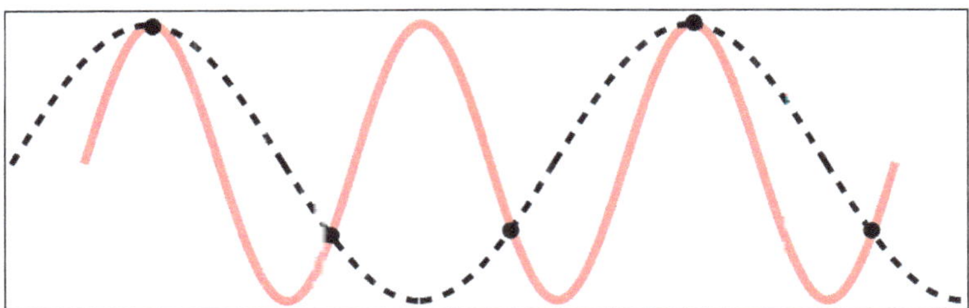

The samples of two sine waves can be identical when at least one of them is at a frequency above half the sample rate.

the Poisson summation formula indicates that the samples, $x(nT)$, of $x(t)$ are sufficient to create a periodic summation of $X(f)$. The result is:

$$X_s(f) \triangleq \sum_{k=-\infty}^{\infty} X(f - kf_s) = \sum_{n=-\infty}^{\infty} T \cdot x(nT)\, e^{-i2\pi nTf},$$

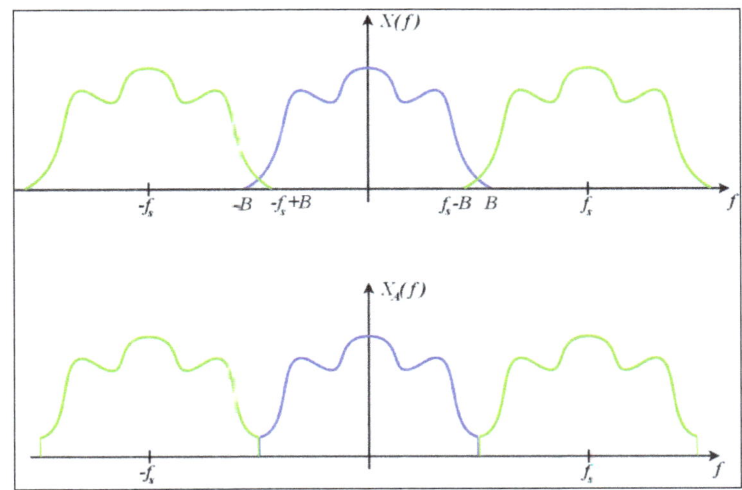

$X(f)$ (top blue) and $X_A(f)$ (bottom blue) are continuous Fourier transforms of two *different* functions, $x(t)$ and $x_A(t)$. When the functions are sampled at rate f_s, the images (green) are added to the original transforms (blue) when one examines the discrete-time Fourier transforms (DTFT) of the sequences. In this hypothetical example, the DTFTs are identical, which means *the sampled sequences are identical*, even though the original continuous pre-sampled functions are not. If these were audio signals, $x(t)$ and $x_A(t)$ might not sound the same. But their samples (taken at rate f_s) are identical and would lead to identical reproduced sounds; thus $x_A(t)$ is an alias of $x(t)$ at this sample rate.

which is a periodic function and its equivalent representation as a Fourier series, whose coefficients are $T \cdot x(nT)$. This function is also known as the discrete-time Fourier transform (DTFT) of the sample sequence.

As depicted, copies of $X(f)$ are shifted by multiples of f_s and combined by addition. For a band-limited function $(X(f) = 0, \text{ for all } |f| \geq B)$ and sufficiently large f_s it is possible for the copies to remain distinct from each other. But if the Nyquist criterion is not satisfied, adjacent copies overlap, and it is not possible in general to discern an unambiguous $X(f)$ Any frequency component above $f_s/2$ is indistinguishable from a lower-frequency component, called an *alias*, associated with one of the copies. In such cases, the customary interpolation techniques produce the alias, rather than the original component. When the sample-rate is pre-determined by other considerations (such as an industry standard), $x(t)$ is usually filtered to reduce its high frequencies to acceptable levels before it is sampled. The type of filter required is a lowpass filter, and in this application it is called an anti-aliasing filter.

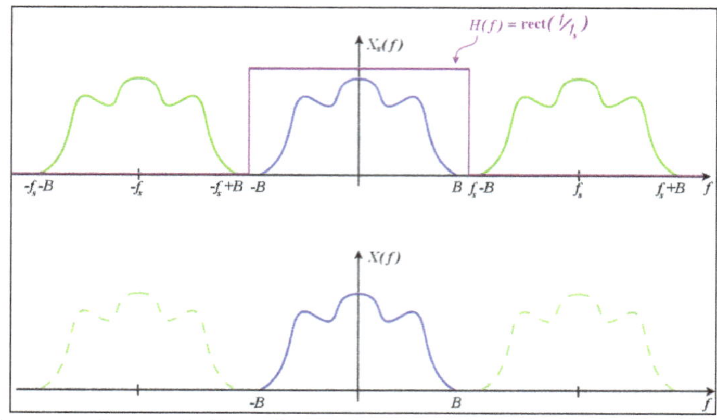

Spectrum, $X_s(f)$, of a properly sampled bandlimited signal (blue) and the adjacent DTFT images (green) that do not overlap. A *brick-wall* low-pass filter, $H(f)$, removes the images, leaves the original spectrum, $X(f)$, and recovers the original signal from its samples.

Derivation as a Special Case of Poisson Summation

When there is no overlap of the copies of $X(f)$, the $k=0$ term of can be recovered by the product:

$$X(f) = H(f) \cdot X_s(f),$$

where,

$$H(f) \square \begin{cases} 1 & |f| < B \\ 0 & |f| > f_s - B. \end{cases}$$

The sampling theorem is proved since $X(f)$ uniquely determines $x(t)$.

All that remains is to derive the formula for reconstruction. $H(f)$ need not be precisely defined in the region $[B, f_s - B]$ because $X_s(f)$ is zero in that region. However, the worst case is when $B = f_s / 2$, the Nyquist frequency. A function that is sufficient for that and all less severe cases is:

$$H(f) = \text{rect}\left(\frac{f}{f_s}\right) = \begin{cases} 1 & |f| < \dfrac{f_s}{2} \\ 0 & |f| > \dfrac{f_s}{2}, \end{cases}$$

where rect(•) is the rectangular function. Therefore:

$$X(f) = \text{rect}\left(\frac{f}{f_s}\right) \cdot X_s(f)$$

$$= \text{rect}(Tf) \cdot \sum_{n=-\infty}^{\infty} T \cdot x(nT)\, e^{-i2\pi nTf}$$

$$= \sum_{n=-\infty}^{\infty} x(nT) \cdot \underbrace{T \cdot \text{rect}(Tf) \cdot e^{-i2\pi nTf}}_{\mathcal{F}\left\{\text{sinc}\left(\frac{t-nT}{T}\right)\right\}}.$$

The inverse transform of both sides produces the Whittaker–Shannon interpolation formula:

$$x(t) = \sum_{n=-\infty}^{\infty} x(nT) \cdot \text{sinc}\left(\frac{t-nT}{T}\right),$$

which shows how the samples, $x(nT)$, can be combined to reconstruct $x(t)$.

- Larger-than-necessary values of f_s (smaller values of T), called *oversampling*, have no effect on the outcome of the reconstruction and have the benefit of leaving room for a *transition band* in which $H(f)$ is free to take intermediate values. Undersampling, which causes aliasing, is not in general a reversible operation.
- Theoretically, the interpolation formula can be implemented as a low pass filter, whose impulse response is $\text{sinc}(t/T)$ and whose input is $\sum_{n=-\infty}^{\infty} x(nT) \cdot \delta(t-nT)$, which is a Dirac comb function modulated by the signal samples. Practical digital-to-analog converters (DAC) implement an approximation like the zero-order hold. In that case, oversampling can reduce the approximation error.

Shannon's original Proof

Poisson shows that the Fourier series produces the periodic summation of $X(f)$, regardless of f and B. Shannon, however, only derives the series coefficients for the case $f_s = 2B$. Virtually quoting Shannon's original paper:

Let $X(\omega)$ be the spectrum of $x(t)$. Then

$$x(t) = \frac{1}{2\pi}\int_{-\infty}^{\infty} X(\omega)e^{i\omega t}\,d\omega = \frac{1}{2\pi}\int_{-2B\pi}^{2\pi B} X(\omega)e^{i\omega t}\,d\omega,$$

because $(\ )$ is assumed to be zero outside the band $\left|\frac{\omega}{2\pi}\right| < B$. If we let $t = \frac{n}{2B}$, where n is any positive or negative integer, we obtain:

$$x\left(\tfrac{n}{2B}\right) = \frac{1}{2\pi}\int_{-2\pi B}^{2\pi B} X(\omega)e^{i\omega\frac{n}{2B}}\,d\omega.$$

On the left are values of $x(t)$ at the sampling points. The integral on the right will be recognized as essentially[a] the nth coefficient in a Fourier-series expansion of the function $X(\omega)$ taking the interval $-B$ to B as a fundamental period. This means that the values of the samples $x(n/2B)$ determine the Fourier coefficients in the series expansion of $X(\omega)$ Thus they determine $X(\omega)$ since

$X(\omega)$ is zero for frequencies greater than B, and for lower frequencies $X(\omega)$ is determined if its Fourier coefficients are determined. But $X(\omega)$ determines the original function $x(t)$ completely, since a function is determined if its spectrum is known. Therefore the original samples determine the function $x(t)$ completely.

Shannon's proof of the theorem is complete at that point, but he goes on to discuss reconstruction via sinc functions, what we now call the Whittaker–Shannon interpolation formula. He does not derive or prove the properties of the sinc function, but these would have been familiar to engineers reading his works at the time, since the Fourier pair relationship between rect (the rectangular function) and sinc was well known.

Let x_n be the nth sample. Then the function $x(t)$ is represented by:

$$x(t) = \sum_{n=-\infty}^{\infty} x_n \frac{\sin \pi(2Bt - n)}{\pi(2Bt - n)}.$$

As in the other proof, the existence of the Fourier transform of the original signal is assumed, so the proof does not say whether the sampling theorem extends to bandlimited stationary random processes.

Application to Multivariable Signals and Images

The sampling theorem is usually formulated for functions of a single variable. Consequently, the theorem is directly applicable to time-dependent signals and is normally formulated in that context. However, the sampling theorem can be extended in a straightforward way to functions of arbitrarily many variables. Grayscale images, for example, are often represented as two-dimensional arrays (or matrices) of real numbers representing the relative intensities of pixels (picture elements) located at the intersections of row and column sample locations. As a result, images require two independent variables, or indices, to specify each pixel uniquely—one for the row, and one for the column.

Color images typically consist of a composite of three separate grayscale images, one to represent each of the three primary colors—red, green, and blue, or *RGB* for short. Other colorspaces using 3-vectors for colors include HSV, CIELAB, XYZ, etc. Some colorspaces such as cyan, magenta, yellow, and black (CMYK) may represent color by four dimensions. All of these are treated as vector-valued functions over a two-dimensional sampled domain.

Similar to one-dimensional discrete-time signals, images can also suffer from aliasing if the sampling resolution, or pixel density, is inadequate. For example, a digital photograph of a striped shirt with high frequencies (in other words, the distance between the stripes is small), can cause aliasing of the shirt when it is sampled by the camera's image sensor. The aliasing appears as a moiré pattern. The "solution" to higher sampling in the spatial domain for this case would be to move closer to the shirt, use a higher resolution sensor, or to optically blur the image before acquiring it with the sensor.

Another example is shown to the right in the brick patterns. The top image shows the effects when the sampling theorem's condition is not satisfied. When software rescales an image (the same process that creates the thumbnail shown in the lower image) it, in effect, runs the image through a low-pass filter first and then downsamples the image to result in a smaller image that does not exhibit the moiré pattern. The top image is what happens when the image is downsampled without low-pass filtering: aliasing results.

Subsampled image showing a Moiré pattern.

Properly sampled image.

The sampling theorem applies to camera systems, where the scene and lens constitute an analog spatial signal source, and the image sensor is a spatial sampling device. Each of these components is characterized by a modulation transfer function (MTF), representing the precise resolution (spatial bandwidth) available in that component. Effects of aliasing or blurring can occur when the lens MTF and sensor MTF are mismatched. When the optical image which is sampled by the sensor device contains higher spatial frequencies than the sensor, the under sampling acts as a low-pass filter to reduce or eliminate aliasing. When the area of the sampling spot (the size of the pixel sensor) is not large enough to provide sufficient spatial anti-aliasing, a separate anti-aliasing filter (optical low-pass filter) may be included in a camera system to reduce the MTF of the optical image. Instead of requiring an optical filter, the graphics processing unit of smartphone cameras performs digital signal processing to remove aliasing with a digital filter. Digital filters also apply sharpening to amplify the contrast from the lens at high spatial frequencies, which otherwise falls off rapidly at diffraction limits.

The sampling theorem also applies to post-processing digital images, such as to up or down sampling. Effects of aliasing, blurring, and sharpening may be adjusted with digital filtering implemented in software, which necessarily follows the theoretical principles.

Critical Frequency

To illustrate the necessity of $f_s > 2B$, consider the family of sinusoids generated by different values of θ in this formula:

$$x(t) = \frac{\cos(2\pi Bt + \theta)}{\cos(\theta)} = \cos(2\pi Bt) - \sin(2\pi Bt)\tan(\theta), \quad -\pi/2 < \theta < \pi/2.$$

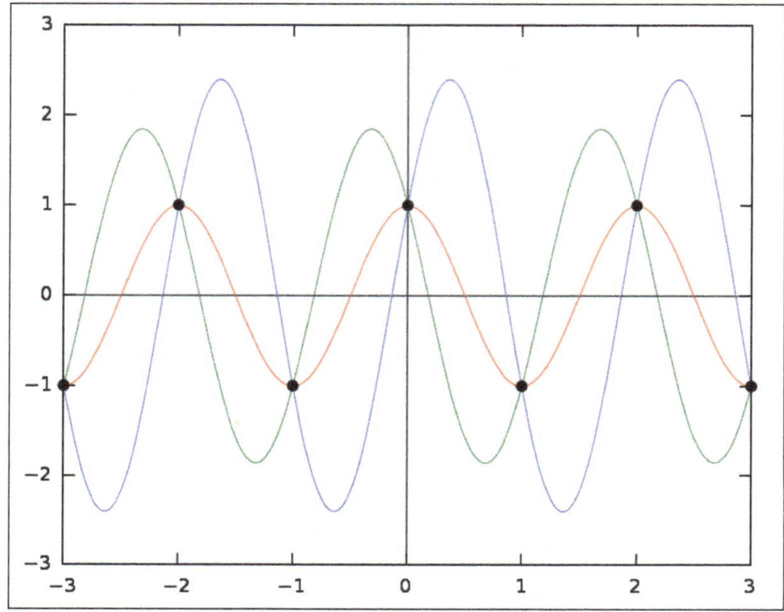

A family of sinusoids at the critical frequency, all having the same sample sequences of alternating +1 and −1. That is, they all are aliases of each other, even though their frequency is not above half the sample rate.

With $f_s = 2B$ or equivalently $T = 1/2B$, the samples are given by:

$$x(nT) = \cos(\pi n) - \underbrace{\sin(\pi n)}_{0}\tan(\theta) = (-1)^n$$

regardless of the value of θ. That sort of ambiguity is the reason for the *strict* inequality of the sampling theorem's condition.

Sampling of Non-baseband Signals

As discussed by Shannon:

> A similar result is true if the band does not start at zero frequency but at some higher value, and can be proved by a linear translation (corresponding physically to single-sideband modulation) of the zero-frequency case. In this case the elementary pulse is obtained from sin(*x*)/*x* by single-side-band modulation.

That is, a sufficient no-loss condition for sampling signals that do not have baseband components exists that involves the *width* of the non-zero frequency interval as opposed to its highest frequency component.

For example, in order to sample the FM radio signals in the frequency range of 100–102 MHz, it is not necessary to sample at 204 MHz (twice the upper frequency), but rather it is sufficient to sample at 4 MHz (twice the width of the frequency interval).

A bandpass condition is that $X(f) = 0$, for all nonnegative f outside the open band of frequencies:

$$\left(\frac{N}{2}f_s, \frac{N+1}{2}f_s\right),$$

for some nonnegative integer N. This formulation includes the normal baseband condition as the case N=0.

The corresponding interpolation function is the impulse response of an ideal brick-wall bandpass filter (as opposed to the ideal brick-wall lowpass filter used above) with cutoffs at the upper and lower edges of the specified band, which is the difference between a pair of lowpass impulse responses:

$$(N+1)\mathrm{sinc}\left(\frac{(N+1)t}{T}\right) - N\mathrm{sinc}\left(\frac{Nt}{T}\right).$$

Other generalizations, for example to signals occupying multiple non-contiguous bands, are possible as well. Even the most generalized form of the sampling theorem does not have a provably true converse. That is, one cannot conclude that information is necessarily lost just because the conditions of the sampling theorem are not satisfied; from an engineering perspective, however, it is generally safe to assume that if the sampling theorem is not satisfied then information will most likely be lost.

Nonuniform Sampling

The sampling theory of Shannon can be generalized for the case of nonuniform sampling, that is, samples not taken equally spaced in time. The Shannon sampling theory for non-uniform sampling states that a band-limited signal can be perfectly reconstructed from its samples if the average sampling rate satisfies the Nyquist condition. Therefore, although uniformly spaced samples may result in easier reconstruction algorithms, it is not a necessary condition for perfect reconstruction.

The general theory for non-baseband and nonuniform samples was developed in 1967 by Henry Landau. He proved that the average sampling rate (uniform or otherwise) must be twice the *occupied* bandwidth of the signal, assuming it is *a priori* known what portion of the spectrum was occupied. In the late 1990s, this work was partially extended to cover signals of when the amount of occupied bandwidth was known, but the actual occupied portion of the spectrum was unknown. In the 2000s, a complete theory was developed using compressed sensing. In particular, the theory, using signal processing language, They show, among other things, that if the frequency locations are unknown, then it is necessary to sample at least at twice the Nyquist criteria; in other words, you must pay at least a factor of 2 for not knowing the location of the spectrum. Note that minimum sampling requirements do not necessarily guarantee stability.

Sampling below the Nyquist Rate under Additional Restrictions

The Nyquist–Shannon sampling theorem provides a sufficient condition for the sampling and reconstruction of a band-limited signal. When reconstruction is done via the Whittaker–Shannon interpolation formula, the Nyquist criterion is also a necessary condition to avoid aliasing, in the sense that if samples are taken at a slower rate than twice the band limit, then there are some signals that will not be correctly reconstructed. However, if further restrictions are imposed on the signal, then the Nyquist criterion may no longer be a necessary condition.

A non-trivial example of exploiting extra assumptions about the signal is given by the recent field of compressed sensing, which allows for full reconstruction with a sub-Nyquist sampling rate. Specifically, this applies to signals that are sparse (or compressible) in some domain. As an example, compressed sensing deals with signals that may have a low over-all bandwidth (say, the *effective* bandwidth EB), but the frequency locations are unknown, rather than all together in a single band, so that the passband technique does not apply. In other words, the frequency spectrum is sparse. Traditionally, the necessary sampling rate is thus $2B$. Using compressed sensing techniques, the signal could be perfectly reconstructed if it is sampled at a rate slightly lower than $2EB$. With this approach, reconstruction is no longer given by a formula, but instead by the solution to a linear optimization program.

Another example where sub-Nyquist sampling is optimal arises under the additional constraint that the samples are quantized in an optimal manner, as in a combined system of sampling and optimal lossy compression. This setting is relevant in cases where the joint effect of sampling and quantization is to be considered, and can provide a lower bound for the minimal reconstruction error that can be attained in sampling and quantizing a random signal. For stationary Gaussian random signals, this lower bound is usually attained at a sub-Nyquist sampling rate, indicating that sub-Nyquist sampling is optimal for this signal model under optimal quantization.

Aliasing

In signal processing and related disciplines, aliasing is an effect that causes different signals to become indistinguishable (or *aliases* of one another) when sampled. It also often refers to the distortion or artifact that results when a signal reconstructed from samples is different from the original continuous signal.

A properly sampled image of a brick wall requires a screen of sufficient resolution to prevent a moiré pattern.

Spatial aliasing in the form of a moiré pattern.

Aliasing can occur in signals sampled in time, for instance digital audio, and is referred to as temporal aliasing. It can also occur in spatially sampled signals; this type of aliasing is called spatial aliasing.

Aliasing is generally avoided by applying low pass filters or anti-aliasing filters (AAF) to the input signal before sampling and when converting a signal from a higher to a lower sampling rate. Suitable reconstruction filtering should then be used when restoring the sampled signal to the continuous domain or converting a signal from a lower to a higher sampling rate. For spatial anti-aliasing, the types of anti-aliasing include fast sample anti-aliasing (FSAA), multisample anti-aliasing, and supersampling.

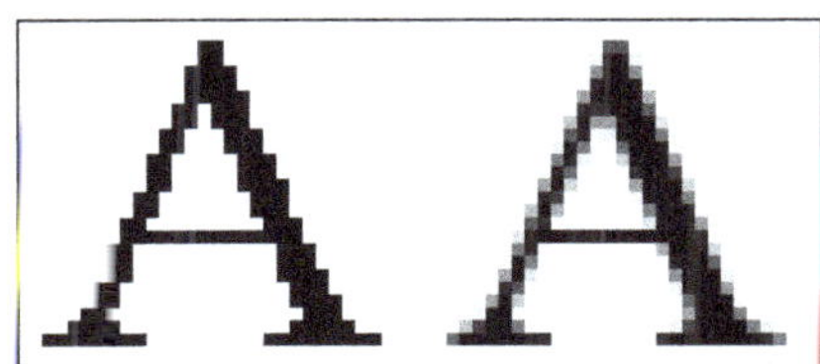

Left: An aliased image of the letter *A* in Times New Roman. Right: An *anti-aliased* image.

When a digital image is viewed, a reconstruction is performed by a display or printer device, and by the eyes and the brain. If the image data is processed in some way during sampling or reconstruction, the reconstructed image will differ from the original image, and an alias is seen.

An example of spatial aliasing is the moiré pattern observed in a poorly pixelized image of a brick wall. Spatial anti-aliasing techniques avoid such poor pixelizations. Aliasing can be caused either by the sampling stage or the reconstruction stage; these may be distinguished by calling sampling aliasing *prealiasing* and reconstruction aliasing *postaliasing*.

Temporal aliasing is a major concern in the sampling of video and audio signals. Music, for instance, may contain high-frequency components that are inaudible to humans. If a piece of music is sampled at 32000 samples per second (Hz), any frequency components at or above 16000 Hz (the Nyquist frequency for this sampling rate) will cause aliasing when the music is reproduced by a digital-to-analog converter (DAC). To prevent this, an anti-aliasing filter is used to remove components above the Nyquist frequency prior to sampling.

In video or cinematography, temporal aliasing results from the limited frame rate, and causes the wagon-wheel effect, whereby a spoked wheel appears to rotate too slowly or even backwards. Aliasing has changed its apparent frequency of rotation. A reversal of direction can be described as a negative frequency. Temporal aliasing frequencies in video and cinematography are determined by the frame rate of the camera, but the relative intensity of the aliased frequencies is determined by the shutter timing (exposure time) or the use of a temporal aliasing reduction filter during filming.

Like the video camera, most sampling schemes are periodic; that is, they have a characteristic sampling frequency in time or in space. Digital cameras provide a certain number of samples (pixels) per degree or per radian, or samples per mm in the focal plane of the camera. Audio signals are sampled (digitized) with an analog-to-digital converter, which produces a constant number of samples per second. Some of the most dramatic and subtle examples of aliasing occur when the signal being sampled also has periodic content.

Bandlimited Functions

Actual signals have a finite duration and their frequency content, as defined by the Fourier transform, has no upper bound. Some amount of aliasing always occurs when such functions are sampled. Functions whose frequency content is bounded (*bandlimited*) have an infinite duration in the time domain. If sampled at a high enough rate, determined by the *bandwidth*, the original function can, in theory, be perfectly reconstructed from the infinite set of samples.

Bandpass Signals

Sometimes aliasing is used intentionally on signals with no low-frequency content, called *bandpass* signals. Undersampling, which creates low-frequency aliases, can produce the same result, with less effort, as frequency-shifting the signal to lower frequencies before sampling at the lower rate. Some digital channelizers exploit aliasing in this way for computational efficiency.

Sampling Sinusoidal Functions

Sinusoids are an important type of periodic function, because realistic signals are often modeled as the summation of many sinusoids of different frequencies and different amplitudes (for example, with a Fourier series or transform). Understanding what aliasing does to the individual sinusoids is useful in understanding what happens to their sum.

When sampling a function at frequency f_s (intervals $1/f_s$), the following functions yield identical sets of samples: $\{\sin(2\pi(f+Nf_s)\,t + \varphi), N = 0, \pm1, \pm2, \pm3, \ldots\}$. A frequency spectrum of the samples produces equally strong responses at all those frequencies. Without collateral information, the frequency of the original function is ambiguous. So the functions and their frequencies are said to be *aliases* of each other. Noting the trigonometric identity:

$$\sin(2\pi(f+Nf_s)t+\phi) = \begin{cases} +\sin(2\pi(f+Nf_s)t+\phi), & f+Nf_s \geq 0 \\ -\sin(2\pi\,|\,f+Nf_s\,|\,t-\phi), & f+Nf_s < 0 \end{cases}$$

we can write all the alias frequencies as positive values: $f_N(f) \triangleq |f+Nf_s|$.

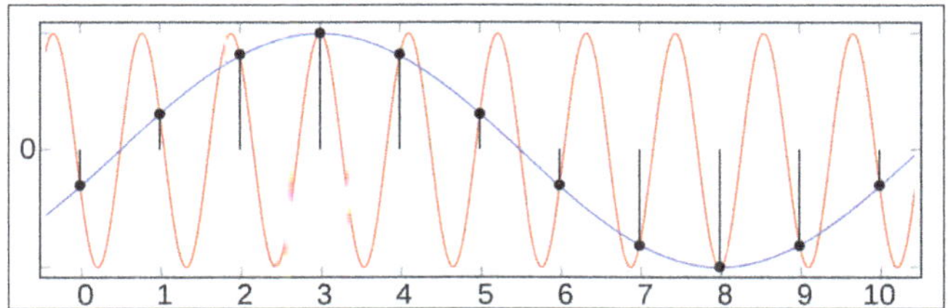
Two different sinusoids that fit the same set of samples.

For example, here a plot depicts a set of samples with parameter $f_s = 1$, and two different sinusoids that could have produced the samples. Nine cycles of the red sinusoid and one cycle of the blue sinusoid span an interval of 10 samples. The corresponding number of *cycles per sample* are $f_{red} = 0.9f_s$ and $f_{blue} = 0.1f_s$. So the $N = -1$ alias of f_{red} is f_{blue} (and vice versa).

Aliasing matters when one attempts to reconstruct the original waveform from its samples. The most common reconstruction technique produces the smallest of the $f_N(f)$ frequencies. So it is usually important that $f_0(f)$ be the unique minimum. A necessary and sufficient condition for that is $f_s/2 > |f|$, where $f_s/2$ is commonly called the Nyquist frequency of a system that samples at rate f_s. In our example, the Nyquist condition is satisfied if the original signal is the blue sinusoid ($f = f_{blue}$). But if $f = f_{red} = 0.9f_s$, the usual reconstruction method will produce the blue sinusoid instead of the red one.

Folding

In the example above, f_{red} and f_{blue} are symmetrical around the frequency $f_s/2$. And in general, as f increases from 0 to $f_s/2$, $f_{-1}(f)$ decreases from f_s to $f_s/2$. Similarly, as f increases from $f_s/2$ to f_s, $f_{-1}(f)$ continues decreasing from $f_s/2$ to 0.

A graph of amplitude vs frequency for a single sinusoid at frequency $0.6 f_s$ and some of its aliases at $0.4 f_s$, $1.4 f_s$, and $1.6 f_s$ would look like the 4 black dots in the first figure below. The red lines depict the paths (loci) of the 4 dots if we were to adjust the frequency and amplitude of the sinusoid along the solid red segment (between $f_s/2$ and f_s). No matter what function we choose to change the amplitude vs frequency, the graph will exhibit symmetry between 0 and f_s. This symmetry is commonly referred to as folding, and another name for $f_s/2$ (the Nyquist frequency) is folding frequency. Folding is often observed in practice when viewing the frequency spectrum of real-valued samples, such as the second figure below.

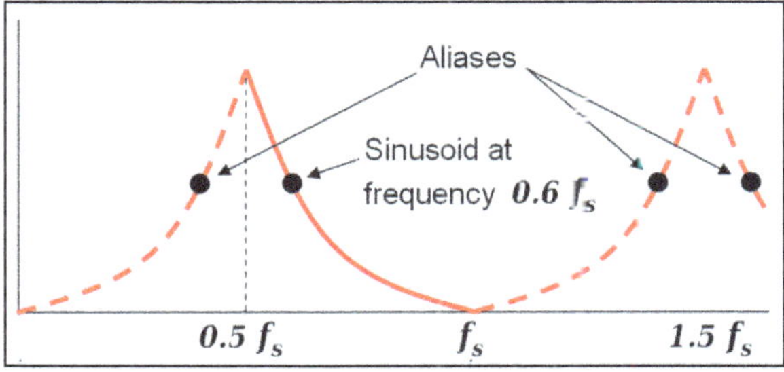

The black dots are aliases of each other. The solid red line is an example of amplitude varying with frequency. The dashed red lines are the corresponding paths of the aliases.

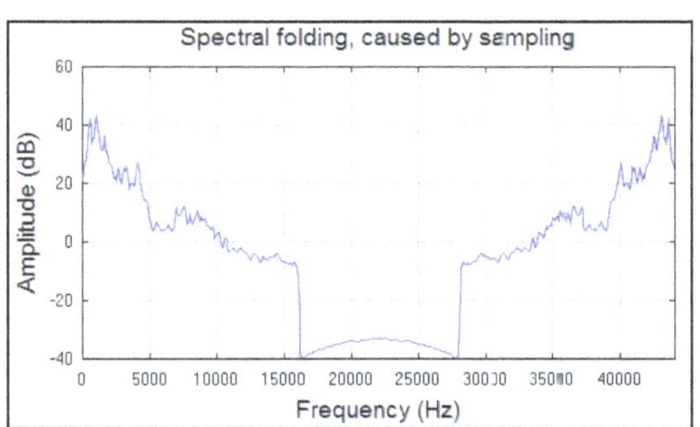

The Fourier transform of music sampled at 44100 samples/sec exhibits symmetry (called "folding") around the Nyquist frequency (22050 Hz).

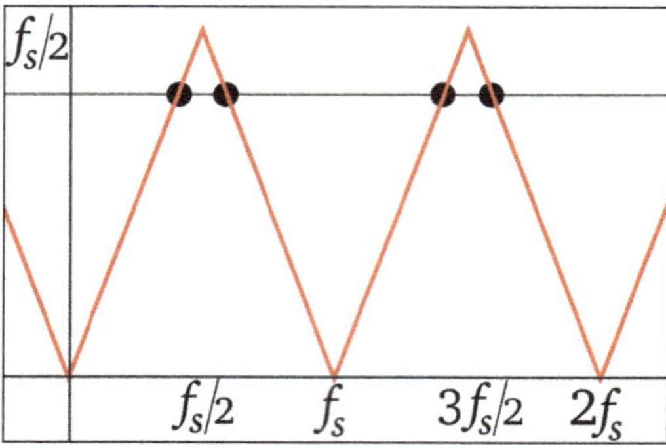

Graph of frequency aliasing, showing folding frequency and periodicity. Frequencies above fs/2 have an alias below fs/2, whose value is given by this graph.

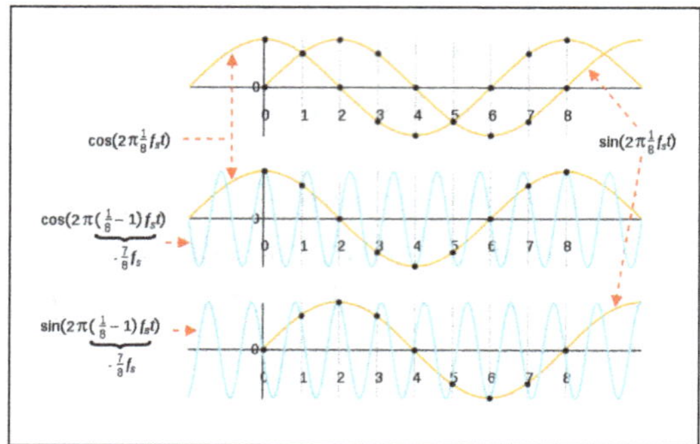

Two complex sinusoids, colored gold and cyan, that fit the same sets of real and imaginary sample points when sampled at the rate (f_s) indicated by the grid lines. The case shown here is: $f_{cyan} = f_{-1}(f_{gold}) = f_{gold} - f_s$.

Complex Sinusoids

Complex sinusoids are waveforms whose samples are complex numbers, and the concept of negative frequency is necessary to distinguish them. In that case, the frequencies of the aliases are given by just: $f_N(f) = f + Nf_s$. Therefore, as f increases from $f_s/2$ to f_s, $f_{-1}(f)$ goes from $-f_s/2$ up to 0. Consequently, complex sinusoids do not exhibit *folding*. Complex samples of real-valued sinusoids have zero-valued imaginary parts and do exhibit folding.

Sample Frequency

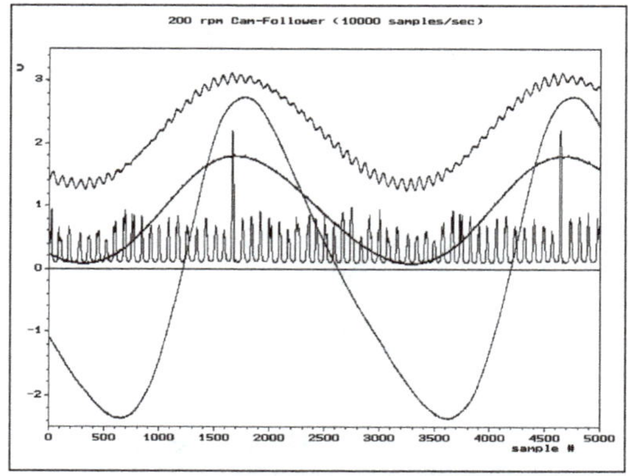

4 waveforms reconstructed from samples taken at six different rates. Two of the waveforms are sufficiently sampled to avoid aliasing at all six rates. The other two illustrate increasing distortion (aliasing) at the lower rates.

When the condition $f_s/2 > f$ is met for the highest frequency component of the original signal, then it is met for all the frequency components, a condition called the Nyquist criterion. That is typically approximated by filtering the original signal to attenuate high frequency components before it is sampled. These attenuated high frequency components still generate low-frequency aliases, but typically at low enough amplitudes that they do not cause problems. A filter chosen in anticipation of a certain sample frequency is called an anti-aliasing filter.

The filtered signal can subsequently be reconstructed, by interpolation algorithms, without significant additional distortion. Most sampled signals are not simply stored and reconstructed. But the fidelity of a theoretical reconstruction (via the Whittaker–Shannon interpolation formula) is a customary measure of the effectiveness of sampling.

Angular Aliasing

Aliasing occurs whenever the use of discrete elements to capture or produce a continuous signal causes frequency ambiguity.

Spatial aliasing, particular of angular frequency, can occur when reproducing a light field or sound field with discrete elements, as in 3D displays or wave field synthesis of sound.

This aliasing is visible in images such as posters with lenticular printing: if they have low angular resolution, then as one moves past them, say from left-to-right, the 2D image does not initially change (so it appears to move left), then as one moves to the next angular image, the image suddenly changes (so it jumps right) – and the frequency and amplitude of this side-to-side movement corresponds to the angular resolution of the image (and, for frequency, the speed of the viewer's lateral movement), which is the angular aliasing of the 4D light field.

The lack of parallax on viewer movement in 2D images and in 3-D film produced by stereoscopic glasses (in 3D films the effect is called "yawing", as the image appears to rotate on its axis) can similarly be seen as loss of angular resolution, all angular frequencies being aliased to 0 (constant).

Online Audio Example

The qualitative effects of aliasing can be heard in the following audio demonstration. Six sawtooth waves are played in succession, with the first two sawtooths having a fundamental frequency of 440 Hz (A4), the second two having fundamental frequency of 880 Hz (A5), and the final two at 1760 Hz (A6). The sawtooths alternate between bandlimited (non-aliased) sawtooths and aliased sawtooths and the sampling rate is 22.05 kHz. The bandlimited sawtooths are synthesized from the sawtooth waveform's Fourier series such that no harmonics above the Nyquist frequency are present.

The aliasing distortion in the lower frequencies is increasingly obvious with higher fundamental frequencies, and while the bandlimited sawtooth is still clear at 1760 Hz, the aliased sawtooth is degraded and harsh with a buzzing audible at frequencies lower than the fundamental.

Direction Finding

A form of spatial aliasing can also occur in antenna arrays or microphone arrays used to estimate the direction of arrival of a wave signal, as in geophysical exploration by seismic waves. Waves must be sampled at more than two points per wavelength, or the wave arrival direction becomes ambiguous.

Anti-Aliasing Filter

An anti-aliasing filter (AAF) is a filter used before a signal sampler to restrict the bandwidth of a signal to approximately or completely satisfy the Nyquist–Shannon sampling theorem over the band of interest. Since the theorem states that unambiguous reconstruction of the signal from its samples is possible when the power of frequencies above the Nyquist frequency is zero, a real anti-aliasing filter trades off between bandwidth and aliasing. A realizable anti-aliasing filter will typically either permit some aliasing to occur or else attenuate some in-band frequencies close to the Nyquist limit. For this reason, many practical systems sample higher than would be theoretically required by a perfect AAF in order to ensure that all frequencies of interest can be reconstructed, a practice called oversampling.

Optical Applications

Simulated photographs of a brick wall without (left) and with (right) an optical low-pass filter.

In the case of optical image sampling, as by image sensors in digital cameras, the anti-aliasing filter is also known as an optical low-pass filter (OLPF), blur filter, or AA filter. The mathematics of sampling in two spatial dimensions is similar to the mathematics of time-domain sampling, but the filter implementation technologies are different. The typical implementation in digital cameras is two layers of birefringent material such as lithium niobate, which spreads each optical point into a cluster of four points.

The choice of spot separation for such a filter involves a tradeoff among sharpness, aliasing, and fill factor (the ratio of the active refracting area of a microlens array to the total contiguous area occupied by the array). In a monochrome or three-CCD or Foveon X3 camera, the microlens array alone, if near 100% effective, can provide a significant anti-aliasing effect, while in color filter array (CFA, e.g. Bayer filter) cameras, an additional filter is generally needed to reduce aliasing to an acceptable level.

The Pentax K-3 from Ricoh introduced a unique sensor-based anti-aliasing filter. The filter works by micro vibrating the sensor element. The user can turn the vibration on or off, selecting anti-aliasing or no anti-aliasing.

Audio Applications

Anti-aliasing filters are commonly used at the input of digital signal processing system's analog to digital converter; similar filters are used as reconstruction filters at the output of such systems, for example in music players. In the latter case, the filter prevents imaging, the reverse process of aliasing where in-band frequencies are mirrored out of band.

Oversampling

A technique known as oversampling is commonly used in audio ADCs. The idea is to use a higher intermediate digital sample rate, so that a nearly ideal digital filter can sharply cut off aliasing near the original low Nyquist frequency and give better phase response, while a much simpler analog filter can stop frequencies above the new higher Nyquist frequency. Because analog filters have relatively high cost and limited performance, relaxing the demands on the analog filter can greatly reduce both aliasing and cost. Furthermore, because some noise is averaged out, the higher sampling rate can moderately improve SNR.

Alternatively, a signal may be intentionally oversampled without an intermediate frequency to reduce the requirements on the anti-alias filter. For example, CD audio typically extends up to 20 kHz, but is sampled with a 22.05 kHz Nyquist rate. By oversampling by 2.05 kHz, both aliasing and attenuation of higher audio frequencies can be prevented even with less than ideal filters.

Bandpass Signals

Often, an anti-aliasing filter is a low-pass filter; this is not a requirement, however. Generalizations of the Nyquist–Shannon sampling theorem allow sampling of other band-limited passband signals instead of baseband signals.

For signals that are bandwidth limited, but not centered at zero, a band-pass filter can be used as an anti-aliasing filter. For example, this could be done with a single-sideband modulated or frequency modulated signal. If one desired to sample an FM radio broadcast centered at 87.9 MHz and bandlimited to a 200 kHz band, then an appropriate anti-alias filter would be centered on 87.9 MHz with 200 kHz bandwidth (or pass-band of 87.8 MHz to 88.0 MHz), and the sampling rate would be no less than 400 kHz, but should also satisfy other constraints to prevent aliasing.

Signal Overload

It is very important to avoid input signal overload when using an anti-aliasing filter. If the signal is strong enough, it can cause clipping at the analog-to-digital converter, even after filtering. When distortion due to clipping occurs after the anti-aliasing filter, it can create components outside the passband of the anti-aliasing filter; these components can then alias, causing the reproduction of other non-harmonically related frequencies.

Signal Quantization

The digitization of analog signals involves the rounding off of the values which are approximately equal to the analog values. The method of sampling chooses a few points on the analog signal and then these points are joined to round off the value to a near stabilized value. Such a process is called as Quantization.

Quantizing an Analog Signal

The analog-to-digital converters perform this type of function to create a series of digital values out of the given analog signal. The following figure represents an analog signal. This signal to get converted into digital, has to undergo sampling and quantizing.

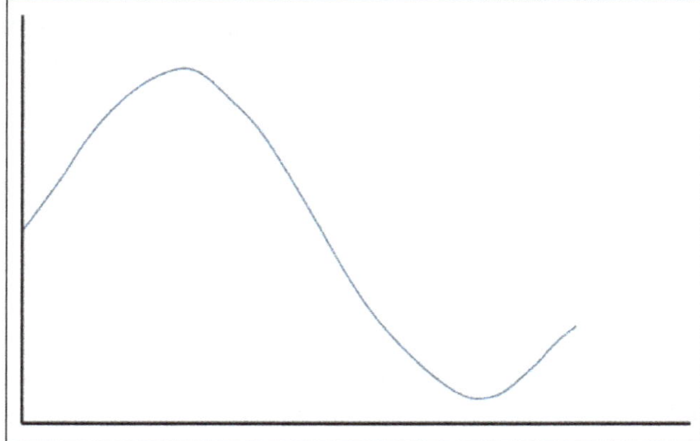

The quantizing of an analog signal is done by discretizing the signal with a number of quantization levels. Quantization is representing the sampled values of the amplitude by a finite set of levels, which means converting a continuous-amplitude sample into a discrete-time signal.

The following figure shows how an analog signal gets quantized. The blue line represents analog signal while the brown one represents the quantized signal.

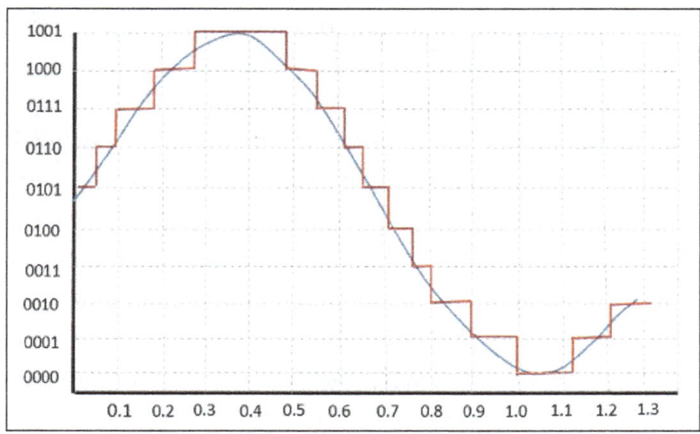

Both sampling and quantization result in the loss of information. The quality of a Quantizer output depends upon the number of quantization levels used. The discrete amplitudes of the quantized output are called as representation levels or reconstruction levels. The spacing between the two adjacent representation levels is called a quantum or step-size.

The following figure shows the resultant quantized signal which is the digital form for the given analog signal.

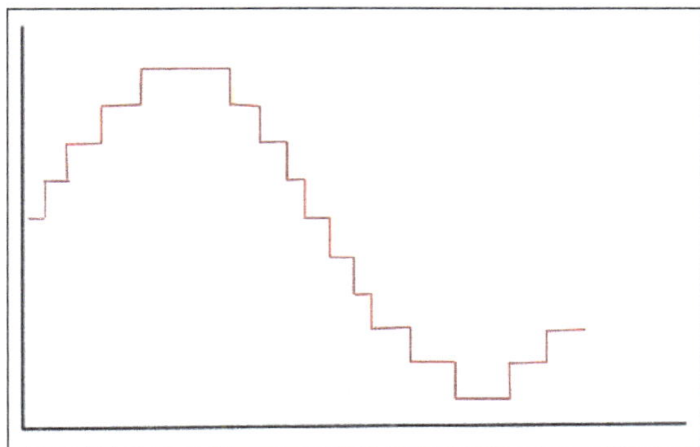

This is also called as Stair-case waveform, in accordance with its shape.

Types of Quantization

There are two types of Quantization - Uniform Quantization and Non-uniform Quantization.

The type of quantization in which the quantization levels are uniformly spaced is termed as a Uniform Quantization. The type of quantization in which the quantization levels are unequal and mostly the relation between them is logarithmic, is termed as a Non-uniform Quantization.

There are two types of uniform quantization. They are Mid-Rise type and Mid-Tread type. The following figures represent the two types of uniform quantization.

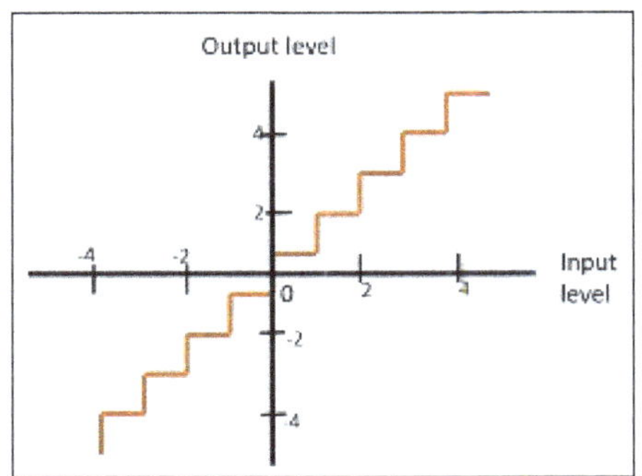

Mid-Rise type Uniform Quantization.

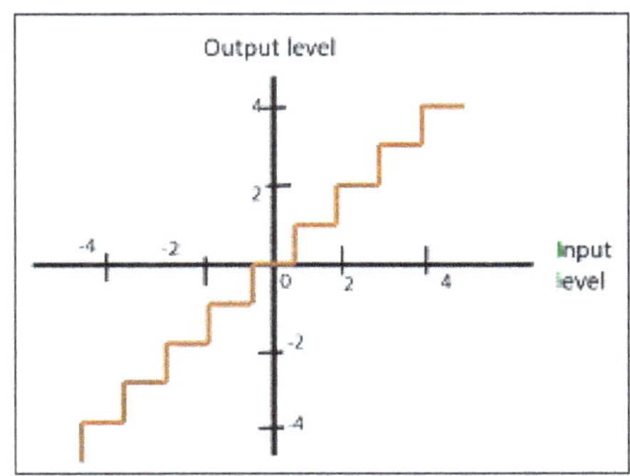

Mid-Tread type Uniform Quantization.

The first figure shown above shows the mid-rise type and second figure shown above shows the mid-tread type of uniform quantization.

- The Mid-Rise type is so called because the origin lies in the middle of a raising part of the stair-case like graph. The quantization levels in this type are even in number.
- The Mid-tread type is so called because the origin lies in the middle of a tread of the stair-case like graph. The quantization levels in this type are odd in number.

- Both the mid-rise and mid-tread type of uniform quantizers are symmetric about the origin.

Quantization Error

For any system, during its functioning, there is always a difference in the values of its input and output. The processing of the system results in an error, which is the difference of those values.

The difference between an input value and its quantized value is called a Quantization Error. A Quantizer is a logarithmic function that performs Quantization (rounding off the value). An analog-to-digital converter (ADC) works as a quantizer.

The following figure illustrates an example for a quantization error, indicating the difference between the original signal and the quantized signal.

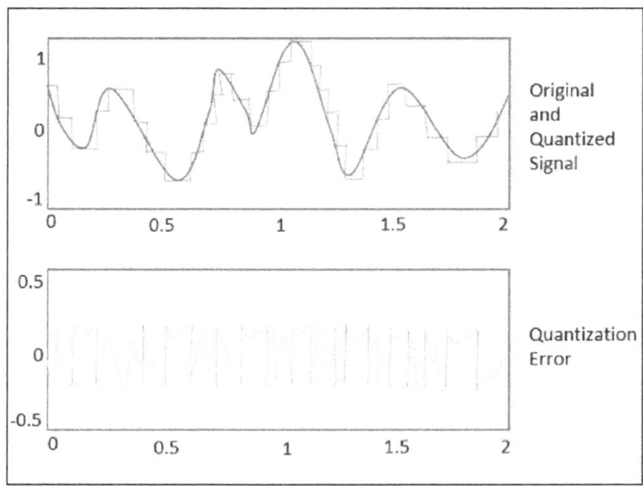

Quantization Noise

It is a type of quantization error, which usually occurs in analog audio signal, while quantizing it to digital. For example, in music, the signals keep changing continuously, where a regularity is not found in errors. Such errors create a wideband noise called as Quantization Noise.

Companding in PCM

The word Companding is a combination of Compressing and Expanding, which means that it does both. This is a non-linear technique used in PCM which compresses the data at the transmitter and expands the same data at the receiver. The effects of noise and crosstalk are reduced by using this technique.

There are two types of Companding techniques.

A-law Companding Technique

- Uniform quantization is achieved at A = 1, where the characteristic curve is linear and no compression is done.

- A-law has mid-rise at the origin. Hence, it contains a non-zero value.
- A-law companding is used for PCM telephone systems.

μ-law Companding Technique

- Uniform quantization is achieved at μ = 0, where the characteristic curve is linear and no compression is done.
- μ-law has mid-tread at the origin. Hence, it contains a zero value.
- μ-law companding is used for speech and music signals.

μ-law is used in North America and Japan.

References

- Signal-processing, computing-dictionaries-thesauruses-pictures-and-press-releases: encyclopedia.com, Retrieved 23 January, 2019
- Stranneby, Dag; Walker, William (2004). Digital Signal Processing and Applications (2nd ed.). Elsevier. ISBN 0-7506-6344-8
- Milić, Ljiljana (2009). Multirate Filtering for Digital Signal Processing. New York: Hershey. p. 192. ISBN 978-1-60566-178-0
- Kipnis, Alon; Goldsmith, Andrea J.; Eldar, Yonina C.; Weissman, Tsachy (January 2016). "Distortion rate function of sub-Nyquist sampled Gaussian sources". IEEE Transactions on Information Theory. 62: 401–429. arXiv:1405.5329. doi:10.1109/tit.2015.2485271
- Digital-communication-quantization, digital-communication: tutorialspoint.com, Retrieved 18 May, 2019

3

Systems used in Signal Processing

A field of electric engineering which aims at analyzing, synthesizing and modifying electromagnetic signals such as of sound, images, videos, etc. is called signal processing. It includes significant systems such as lumped parameter and distributed parameter systems, casual and non-casual systems, linear and non-linear systems and discrete time system. This chapter closely examines these systems used in signal processing to provide an extensive understanding of the subject.

System can be considered as a physical entity which manipulates one or more input signals applied to it. For example a microphone is a system which converts the input acoustic (voice or sound) signal into an electric signal. A system is defined mathematically as a unique operator or transformation that maps an input signal in to an output signal. This is defined as $y(t)\ T\left[x(t)\right]$ where $x(t)$ is input signal, $y(t)$ is output signal, T[] is transformation that characterizes the system behavior.

Lumped Parameter and Distributed Parameter Systems

- A lumped system is one in which the dependent variables of interest are a function of time alone. In general, this will mean solving a set of ordinary differential equations (ODEs).
- A distributed system is one in which all dependent variables are functions of time and one or more spatial variables. In this case, we will be solving partial differential equations (PDEs).

For example, consider the following two systems:

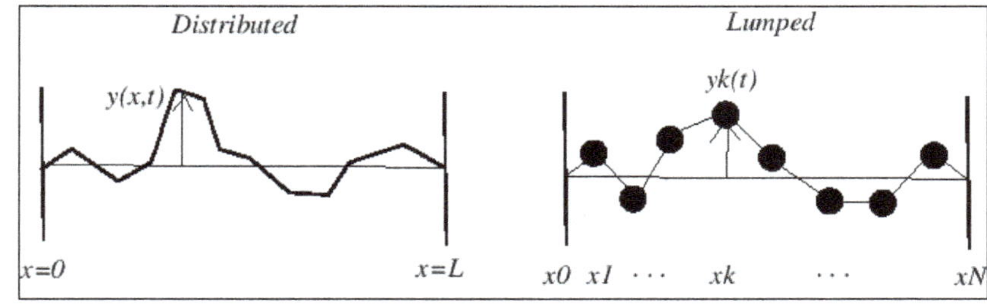

- The first system is a distributed system, consisting of an infinitely thin string, supported at both ends; the dependent variable, the vertical position of the string y(x,t) is indexed continuously in both space and time.
- The second system, a series of beads connected by massless string segments, constrained to move vertically, can be thought of as a lumped system, perhaps an approximation to the continuous string.
- For electrical systems, consider the difference between a lumped RLC network and a transmission line.

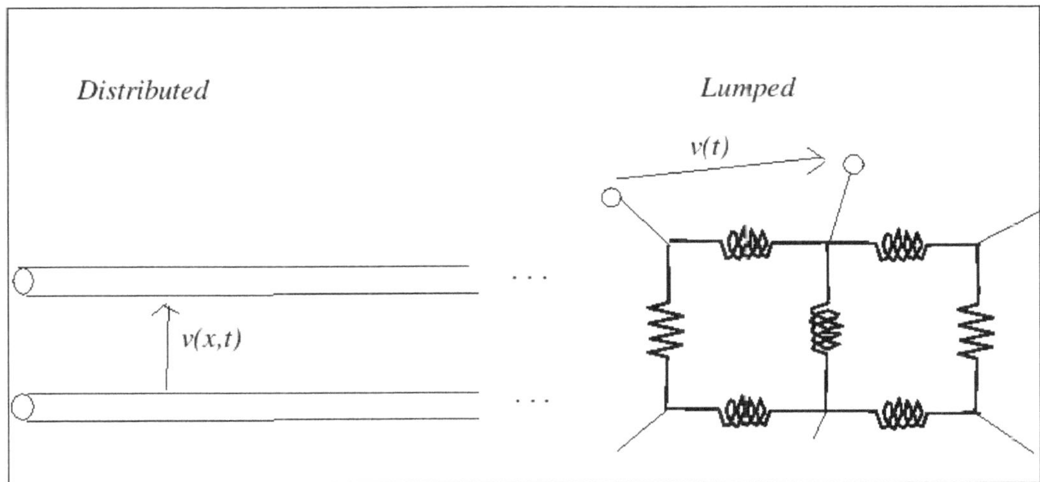

- The importance of lumped approximations to distributed systems will become obvious later, especially for waveguide-based physical modeling, because it enables one to cut computational costs by solving ODEs at a few points, rather than a full PDE (generally much more costly).

Static and Dynamic Systems

Static Systems

Definition: It is a system in which output at any instant of time depends on input sample at the same time.

Example:

$$y(n) = 9x(n)$$

In this example 9 is constant which multiplies input x(n). But output at nth instant that means y(n) depends on the input at the same (nth) time instant x(n). So this is static system.

$$y(n) = x2(n) + 8x(n) + 17$$

Here also output at nth instant, y(n) depends on the input at nth instant. So this is static system.

Why static systems are memory less systems?

Answer:

Observe the input output relations of static system. Output does not depend on delayed $\left[x(n-k)\right]$ or advanced $\left[x(n+k)\right]$ input signals. It only depends on present input (nth) input signal. If output depends upon delayed input signals then such signals should be stored in memory to calculate the output at nth instant. This is not required in static systems. Thus for static systems, memory is not required. Therefore static systems are memory less systems.

Dynamic Systems

It is a system in which output at any instant of time depends on input sample at the same time as well as at other times.

Here other time means, other than the present time instant. It may be past time or future time. Note that if $x(n)$ represents input signal at present instant then,

$x(n-k)$; that means delayed input signal is called as past signal.

$x(n+k)$; that means advanced input signal is called as future signal.

Thus in dynamic systems, output depends on present input as well as past or future inputs.

Examples:

$$y(n) = x(n) + 6x(n-2)$$

Here output at nth instant depends on input at nth instant, $x(n)$ as well as (n-2)th instant x(n-2) is previous sample. So the system is dynamic.

$$y(n) = 4x(n+7) + x(n)$$

Here x(n+7) indicates advanced version of input sample that means it is future sample therefore this is dynamic system.

Why dynamic system has a memory?

Observe input output relations of dynamic system. Since output depends on past or future input sample; we need a memory to store such samples. Thus dynamic system has a memory.

For continuous time (CT) systems:

A continuous time system is static or memoryless if its output depends upon the present input only.

Example:

Voltage drop across a resistor.

It is given by,

$$v(t) = i(t)*R$$

Here the voltage drop depends on the value of the current at that instant. So it is static system.

On the other hand a CT system is dynamic if output depends on present as well as past values.

Causal and Non-causal Systems

Causal Systems

A system is said to be causal system if its output depends on present and past inputs only and not on future inputs.

Examples: The output of casual system depends on present and past inputs, it means $y(n)$ is a function of $x(n-1)$, $x(n-2)$, $x(n-3)$... etc. Some examples of causal systems are given below:

$$y(n) = x(n) + x(n-2)$$

$$y(n) = x(n-1) - x(n-3)$$

$$y(n) = 7x(n-5)$$

Significance of causal systems:

Since causal system does not include future input samples; such system is practically realizable. That mean such system can be implemented practically. Generally all real time systems are causal systems; because in real time applications only present and past samples are present. Since future samples are not present; causal system is memory less system.

Anti Causal or Non-causal System

Definition: A system whose present response depends on future values of the inputs is called as a non-causal system.

Examples: In this case, output $Y(n)$ is function of $x(n)$, $x(n-1)$, $x(n-2)$... etc. as well as it is function of $x(n+1)$, $x(n+2)$, $x(n+3)$, etc. following are some examples of non-causal systems:

$$Y(n) = x(n) + x(n+1)$$

$$Y(n) = 7x(n+2)$$

$$Y(n) = x(n) + 9x(n+5)$$

Significance of Non-causal Systems

Since non-causal system contains future samples; a non-causal system is practically not realizable. That means in practical cases it is not possible to implement a non-causal system.

But if the signals are stored in the memory and at a later time they are used by a system then such signals are treated as advanced or future signal. Because such signals are already present, before the system has started its operation. In such cases it is possible to implement a non-causal system.

Some practical examples of non-causal systems are as follows:

- Population growth,
- Weather forecasting,
- Planning commission etc.

For Continuous Time (C.T.) System

A C.T. system is said to be causal if it produces a response $y(t)$ only after the application of excitation $x(t)$. That means for a causal system the response does not begin before the application of the input $x(t)$.

The other way of defining the causal system is as follows:

A system is said to be causal if its output depends on present and past values of the input and not on the future inputs. If the input is applied at t = tm then the output at t = tm y(tm) will be dependent only on the values of $x(t)$ for t = tm.

Condition for causality: $y(tm) = f[x(t);\ t = tm]$

Causal systems are physically realizable systems. The non-causal systems do not satisfy above condition. Non-causal systems are not physically realizable.

Condition for causality in terms of impulse response $y(t)$:

The relation between $y(t)$ and $x(t)$ is given by:

$$y(t) = x(t) * h(t)$$

Where * represents convolution and $h(t)$ is the impulse response of the system. The condition for causality in terms of the impulse response is as follows:

Condition for causality: $h(t) = 0$ for $t < 0$

This condition states that a linear time invariant (LTI) system is causal if its impulse response $h(t)$ has a zero value for negative values of time.

Solved Problems on Causal and Non-causal System

Determine if the systems described by the following equations are causal or non-causal.

$$y(n) = x(n) + x(n-3)$$

Solution: the given system is causal because its output (y(n)) depends only on the present $x(n)$ and past $x(n-3)$ inputs.

$$y(n) = x(-n+2)$$

Solution: this is non-causal system. This is because at $n = -1$ we get $y(-1) = x[-(-1)+2] = x(3)$. Thus present output at $n = -1$, expects future input i.e. $x(3)$.

Invertible and Non-invertible Systems

A system is said to be invertible if distinct inputs lead to distinct outputs. For such a system there exists an inverse transformation (inverse system) denoted by $T^{-1}[]$ which maps the outputs of original systems to the inputs applied. Accordingly we can write:

$$TT^{-1} = T^{-1}T = I$$

Where $I = 1$ one for single input and single output systems.

A non-invertible system is one in which distinct inputs leads to same outputs. For such a system an inverse system will not exist.

Linear and Non-linear Systems

A linear system is a system which follows the superposition principle. Let us consider a system having its response as 'T', input as $x(n)$ and it produces output $y(n)$.

Let us consider two inputs. Input $x1(n)$ produces output $y1(n)$ and input $x2(n)$ produces output $y2(n)$. Now consider two arbitrary constants $a1$ and $a2$. Then simply multiply these constants with input $x1(n)$ and $x2(n)$ respectively. Thus $a1x1(n)$ produces output $a1y1(n)$ and $a2x2(n)$ produces output $a2y2(n)$.

Theorem for Linearity of the System

A system is said to be linear if the combined response of a1x1(n) and a2x2(n) is equal to the addition of the individual responses.

That means:

$$T[a1\ x1(n) + a2\ x2(n)] = a1\ T[x1(n)] + a2\ T[x2(n)] \ldots\ldots\ldots\ldots\ldots 1)$$

The above theorem is also known as superposition theorem.

Important Characteristic:

Linear system has one important characteristic: If the input to the system is zero then it produces zero output. If the given system produces some output (non-zero) at zero input then the system is said to be Non-linear system. If this condition is satisfied then apply the superposition theorem to determine whether the given system is linear or not?

For continuous time system:

Similar to the discrete time system a continuous time system is said to be linear if it follows the superposition theorem.

Let us consider two systems as follows:

$$y1(t) = f[x1(t)]$$

And $y2(t)=f[x2(t)]$

Here $y1(t)$ and $y2(t)$ are the responses of the system and $x1(t)$ and $x2(t)$ are the excitations. Then the system is said to be linear if it satisfies the following expression:

$$f[a1\ x1(t)+a2\ x2(t)]=a1\ y1(t)+a2\ y2(t)$$

Where $a1$ and $a2$ are constants.

A system is said to be non-linear system if does not satisfies the above expression. Communication channels and filters are examples of linear systems.

How to determine whether the given system is Linear or not?

To determine whether the given system is Linear or not, we have to follow the following steps:

Step 1: Apply zero input and check the output. If the output is zero then the system is linear. If this step is satisfied then follow the remaining steps.

Step 2: Apply individual inputs to the system and determine corresponding outputs. Then add all outputs. Denote this addition by $y'(n)$. This is the R.H.S. of the 1st equation.

Step 3: Combine all inputs. Apply it to the system and find out y"(n). This is L.H.S. of equation.

Step 4: if $y'(n)=y''(n)$ then the system is linear otherwise it is non-linear system.

Example:

Determine whether the following system is linear or not.

$$y(n) = n\,x(n)$$

Solution:

Step 1: When input x(n) is zero then output is also zero. Here first step is satisfied so we will check remaining steps for linearity.

Step 2: Let us consider two inputs $x1(n)$ and $x2(n)$ be the two inputs which produces outputs $y1(t)$ and $y2(t)$ respectively. It is given as follows:

$$x1(n) \xrightarrow{T} y1(n) = n\,x1(n)$$

$$\text{And } x2(n) \xrightarrow{T} y2(n) = n\,x2(n)$$

Now add these two output to get $y'(n)$

Therefore $y'(n) = y1(n) + y2(n) + n\,x1(n) + n\,x2(n)$

Therefore $y'(n) = n\left[x1(n) + x2(n)\right]$

Step 3: Now add $x1(n)$ and $x2(n)$ and apply this input to the system.

Therefore:

$$\left[x1(n) + x2(n)\right] \xrightarrow{T} y''(n) = T\left[x1(n) + x2(n)\right] = n\left[x1(n) + x2(n)\right]$$

We know that the function of system is to multiply input by 'n'.

Here $\left[x1(n) + x2(n)\right]$ acts as one input to the system. So the corresponding output is:

$$y''(n) = n\left[x1(n) + x2(n)\right]$$

Step 4: Compare $y'(n)$ and $y''(n)$.

Here $y'(n) = y''(n)$. hence the given system is linear.

Stable and Unstable Systems

To define stability of a system we will use the term 'BIBO'. It stands for Bounded Input Bounded Output. The meaning of word 'bounded' is some finite value. So bounded input means input signal is having some finite value. i.e. input signal is not infinite. Similarly bounded output means, the output signal attains some finite value i.e. the output is not reaching to infinite level.

An infinite system is BIBO stable if and only if every bounded input produces bounded output.

Mathematical representation:

Let us consider some finite number Mx whose value is less than infinite. That means $Mx < 8$, so it's a finite value. Then if input is bounded, we can write:

$$|x(n)| = Mx < 8$$

Similarly for C.T. system:

$$|x(t)| = Mx < 8$$

Similarly consider some finite number My whose value is less than infinity. That means $My < 8$, so it's a finite value. Then if output is bounded, we can write:

$$|y(n)| = My < 8$$

Similarly for continuous time system:

$$|y(t)| = My < 8$$

An initially system is said to be unstable if bounded input produces unbounded (infinite) output.

Significance

- Unstable system shows erratic and extreme behavior.
- When unstable system is practically implemented then it causes overflow.

Solved Problem on Stability

Determine whether the following discrete time functions are stable or not.

$$y(n) = x(-n)$$

Solution: we $x(-n)$ (-n) should be finite. So when input is bounded output will be bounded. Thus the given function is Stable system.

Continuous-time System

A continuous time system operates on a continuous time signal input and produces a continuous time signal output. There are numerous examples of useful continuous time systems in signal processing as they essentially describe the world around us. The class of continuous time systems that are both linear and time invariant, known as continuous time LTI systems, is of particular interest as the properties of linearity and time invariance together allow the use of some of the most important and powerful tools in signal processing.

Linearity and Time Invariance

A system H is said to be linear if it satisfies two important conditions. The first, additivity, states for every pair of signals x, y that $H(x+y) = H(x) + H(y)$ The second, homogeneity of degree one, states for every signal x and scalar a we have $H(ax) = aH(x)$. It is clear that these conditions can be combined together into a single condition for linearity. Thus, a system is said to be linear if for every signals x, y and scalars a, b we have that:

$$H(ax+by)=aH(x)+bH(y)$$

Linearity is a particularly important property of systems as it allows us to leverage the powerful tools of linear algebra, such as bases, eigenvectors, and eigenvalues, in their study.

A system H is said to be time invariant if a time shift of an input produces the corresponding shifted output. In other, more precise words, the system H commutes with the time shift operator S_T for every $T \in R$. That is:

$$S_T H = HS_T.$$

Time invariance is desirable because it eases computation while mirroring our intuition that, all else equal, physical systems should react the same to identical inputs at different times.

When a system exhibits both of these important properties it allows for a more straigtforward analysis than would otherwise be possible. As will be explained and proven in subsequent modules, computation of the system output for a given input becomes a simple matter of convolving the input with the system's impulse response signal. Also proven later, the fact that complex exponential are eigenvectors of linear time invariant systems will enable the use of frequency domain tools such as the various Fouier transforms and associated transfer functions, to describe the behavior of linear time invariant systems.

Example:

Consider the system H in which:

$$H(f(t))=2f(t)$$

for all signals f. Given any two signals f, g and scalars a, b

$$H(af(t)+bg(t)))=2(af(t)+bg(t))=a2f(t)+b2g(t)=aH(f(t))+bH(g(t))$$

for all real t. Thus, H is a linear system. For all real T and signals f,

$$ST(H(f(t)))=ST(2f(t))=2f(t-T)=H(f(t-T))=H(ST(f(t)))$$

for all real . Thus, H is a time invariant system. Therefore, H is a linear time invariant system.

Differential Equation Representation

It is often useful to to describe systems using equations involving the rate of change in some quantity. For continuous time systems, such equations are called differential equations. One important class of differential equations is the set of linear constant coefficient ordinary differential equations.

Example:

Consider the series RLC circuit shown in figure. This system can be modeled using differential equations. We can use the voltage equations for each circuit element and Kirchoff's voltage law to write a second order linear constant coefficient differential equation describing the charge on the capacitor.

The voltage across the battery is simply V. The voltage across the capacitor is $\frac{1}{C}$q. The voltage across the resistor is $R\frac{dq}{dt}$. Finally, the voltage across the inductor is $L\frac{d^2q}{dt^2}$. Therefore, by Kirchoff's voltage law, it follows that:

$$L\frac{d^2q}{dt^2}+R\frac{dq}{dt}+\frac{1}{C}q=V.$$

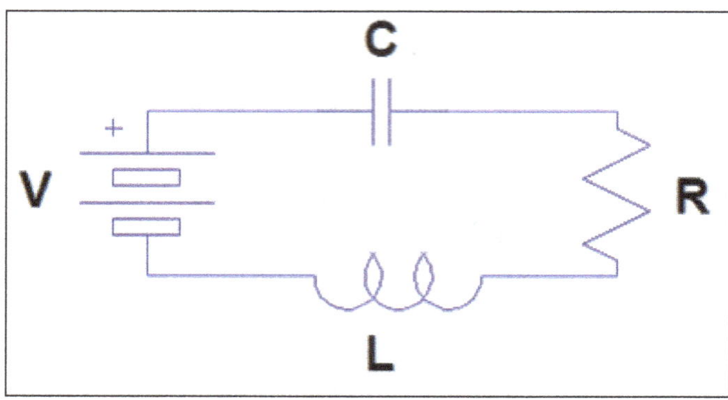

A series RLC circuit.

Discrete Time System

A discrete time system operates on a discrete time signal input and produces a discrete time signal output. There are numerous examples of useful discrete time systems in digital signal processing, such as digital filters for images or sound. The class of discrete time systems that are both linear and time invariant, known as discrete time LTI systems, is of particular interest as the properties of linearity and time invariance together allow the use of some of the most important and powerful tools in signal processing.

Linearity and Time Invariance

A system H is said to be linear if it satisfies two important conditions. The first, additivity, states for every pair of signals x, y that $H(x+y)=H(x)+H(y)$. The second, homogeneity of degree one, states for every signal x and scalar a we have $H(ax)=aH(x)$. It is clear that these conditions can be combined together into a single condition for linearity. Thus, a system is said to be linear if for every signals x, y and scalars a, b we have that:

$$H(ax+by)=aH(x)+bH(y).$$

Linearity is a particularly important property of systems as it allows us to leverage the powerful tools of linear algebra, such as bases, eigenvectors, and eigenvalues, in their study.

A system H is said to be time invariant if a time shift of an input produces the corresponding shifted output. In other, more precise words, the system H commutes with the time shift operator S_T for every $T \in Z$. That is:

$$S_T H = H S_T.$$

Time invariance is desirable because it eases computation while mirroring our intuition that, all else equal, physical systems should react the same to identical inputs at different times.

When a system exhibits both of these important properties it opens. As will be explained and proven in subsequent modules, computation of the system output for a given input becomes a simple matter of convolving the input with the system's impulse response signal. the fact that complex exponential are eigenvectors of linear time invariant systems will encourage the use of frequency domain tools such as the various Fouier transforms and associated transfer functions, to describe the behavior of linear time invariant systems.

Example:

Consider the system H in which:

$$H(f(n)) = 2f(n)$$

for all signals f. Given any two signals f, g and scalars a b

$$H(af(n) + bg(n))) = 2(af(n) + bg(n)) = a2f(n) + b2g(n) = aH(f(n)) + bH(g(n))$$

for all integers n. Thus, H is a linear system. For all integers T and signals f,

$$S_T(H(f(n))) = S_T(2f(n)) = 2f(n-T) = H(f(n-T)) = H(S_T(f(n)))$$

for all integers n. Thus, H is a time invariant system. Therefore, H is a linear time invariant system.

Difference Equation Representation

It is often useful to to describe systems using equations involving the rate of change in some quantity. For discrete time systems, such equations are called difference equations, a type of recurrence relation. One important class of difference equations is the set of linear constant coefficient difference equations.

Example:

Recall that the Fibonacci sequence describes a (very unrealistic) model of what happens when a pair rabbits get left alone in a black box. The assumptions are that a pair of rabits never die and produce a pair of offspring every month starting on their second month of life. This system is defined by the recursion relation for the number of rabit pairs $y(n)$ at month n:

$$y(n) = y(n-1) + y(n-2)$$

with the initial conditions $y(0) = 0$ and $y(1) = 1$. The result is a very fast growth in the sequence. This is why we never leave black boxes open.

Finite Impulse Response System

In signal processing, a **finite impulse response (FIR) filter** is a filter whose impulse response (or response to any finite length input) is of *finite* duration, because it settles to zero in finite time. This is in contrast to infinite impulse response (IIR) filters, which may have internal feedback and may continue to respond indefinitely (usually decaying).

The impulse response (that is, the output in response to a Kronecker delta input) of an Nth-order discrete-time FIR filter lasts exactly $N + 1$ samples (from first nonzero element through last nonzero element) before it then settles to zero.

FIR filters can be discrete-time or continuous-time, and digital or analog.

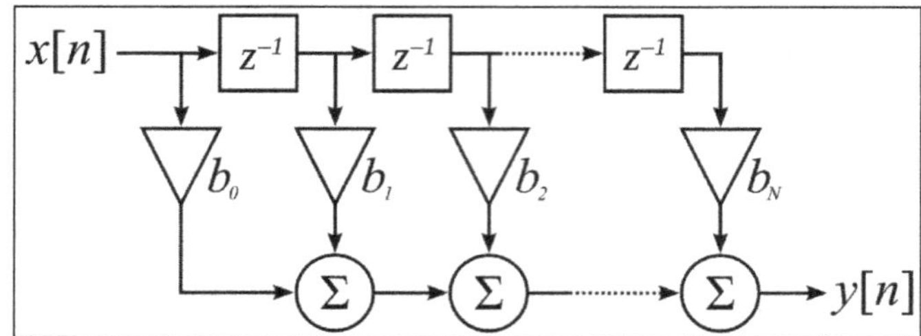

A direct form discrete-time FIR filter of order *N*. The top part is an *N*-stage delay line with $N + 1$ taps. Each unit delay is a z^{-1} operator in Z-transform notation.

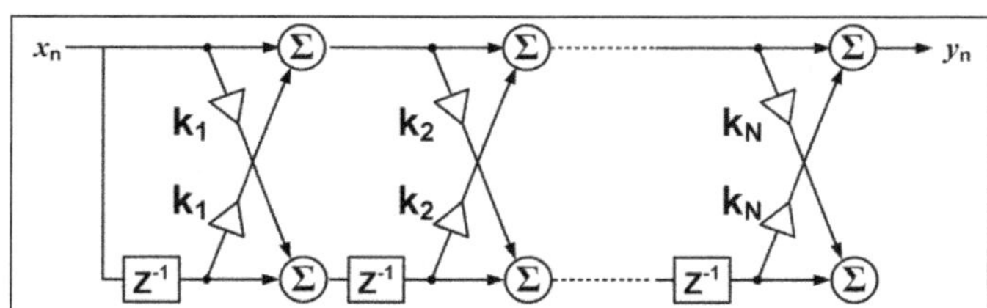

A lattice-form discrete-time FIR filter of order *N*. Each unit delay is a z^{-1} operator in Z-transform notation.

For a causal discrete-time FIR filter of order *N*, each value of the output sequence is a weighted sum of the most recent input values:

$$
\begin{aligned}
y[n] &= b_0 x[n] + b_1 x[n-1] + \cdots + b_N x[n-N] \\
&= \sum_{i=0}^{N} b_i \cdot x[n-i],
\end{aligned}
$$

where:

- $x[n]$ is the input signal,
- $y[n]$ is the output signal,
- N is the filter order; an Nth-order filter has $(N+1)$ terms on the right-hand side,

- b_i is the value of the impulse response at the i'th instant for $0 \le i \le N$ of an Nth-order FIR filter. If the filter is a direct form FIR filter then b_i is also a coefficient of the filter.

This computation is also known as discrete convolution.

The $x[n-i]$ in these terms are commonly referred to as *taps*, based on the structure of a tapped delay line that in many implementations or block diagrams provides the delayed inputs to the multiplication operations. One may speak of a 5th order/6-tap filter, for instance.

The impulse response of the filter as defined is nonzero over a finite duration. Including zeros, the impulse response is the infinite sequence:

$$h[n] = \sum_{i=0}^{N} b_i \cdot \delta[n-i] = \begin{cases} b_n & 0 \le n \le N \\ 0 & \text{otherwise.} \end{cases}$$

If an FIR filter is non-causal, the range of nonzero values in its impulse response can start before $n = 0$, with the defining formula appropriately generalized.

Properties

An FIR filter has a number of useful properties which sometimes make it preferable to an infinite impulse response (IIR) filter. FIR filters:

- Require no feedback. This means that any rounding errors are not compounded by summed iterations. The same relative error occurs in each calculation. This also makes implementation simpler.
- Are inherently stable, since the output is a sum of a finite number of finite multiples of the input values, so can be no greater than $\sum |b_i|$ times the largest value appearing in the input.
- Can easily be designed to be linear phase by making the coefficient sequence symmetric. This property is sometimes desired for phase-sensitive applications, for example data communications, seismology, crossover filters, and mastering.

The main disadvantage of FIR filters is that considerably more computation power in a general purpose processor is required compared to an IIR filter with similar sharpness or selectivity, especially when low frequency (relative to the sample rate) cutoffs are needed. However, many digital signal processors provide specialized hardware features to make FIR filters approximately as efficient as IIR for many applications.

Frequency Response

The filter's effect on the sequence $x[n]$ is described in the frequency domain by the convolution theorem:

$$\underbrace{\mathcal{F}\{x*h\}}_{Y(\omega)} = \underbrace{\mathcal{F}\{x\}}_{X(\omega)} \cdot \underbrace{\mathcal{F}\{h\}}_{H(\omega)} \text{ and } y[n] = x[n]*h[n] = \mathcal{F}^{-1}\{X(\omega)\cdot H(\omega)\},$$

where operators $\mathcal{F}$ and $\mathcal{F}^{-1}$ respectively denote the discrete-time Fourier transform (DTFT) and its inverse. Therefore, the complex-valued, multiplicative function $H(\omega)$ is the filter's frequency response. It is defined by a Fourier series:

$$H_{2\pi}(\omega) \triangleq \sum_{n=-\infty}^{\infty} h[n]\cdot\left(e^{i\omega}\right)^{-n} = \sum_{n=0}^{N} b_n\cdot\left(e^{i\omega}\right)^{-n},$$

where the added subscript denotes 2π-periodicity. Here ω represents frequency in normalized units (*radians/sample*). The substitution $\omega = 2\pi f$, favored by many filter design programs, changes the units of frequency (f) to *cycles/sample* and the periodicity to 1. When the x[n] sequence has a known sampling-rate, f_s *samples/second*, the substitution $\omega = 2\pi f / f_s$ changes the units of frequency (f) to *cycles/second* (hertz) and the periodicity to f_s The value $\omega = \pi$ corresponds to a frequency of $f = \frac{f_s}{2}\, Hz = \frac{1}{2}$ *cycles/sample*, which is the Nyquist frequency.

$H_{2\pi}(\omega)$ can also be expressed in terms of the Z-transform of the filter impulse response:

$$\hat{H}(z) \triangleq \sum_{n=-\infty}^{\infty} h[n]\cdot z^{-n}.$$

$$H_{2\pi}(\omega) = \hat{H}(z)\Big|_{z=e^{j\omega}} = \hat{H}(e^{j\omega}).$$

Filter Design

An FIR filter is designed by finding the coefficients and filter order that meet certain specifications, which can be in the time domain (e.g. a matched filter) and/or the frequency domain (most common). Matched filters perform a cross-correlation between the input signal and a known pulse shape. The FIR convolution is a cross-correlation between the input signal and a time-reversed copy of the impulse response. Therefore, the matched filter's impulse response is designed by sampling the known pulse-shape and using those samples in reverse order as the coefficients of the filter.

When a particular frequency response is desired, several different design methods are common:

- Window design method.
- Frequency Sampling method.
- Weighted least squares design.
- Parks-McClellan method (also known as the Equiripple, Optimal, or Minimax method). The Remez exchange algorithm is commonly used to find an optimal equiripple set of coefficients. Here the user specifies a desired frequency response, a weighting function for errors from this response, and a filter order *N*. The algorithm then finds the set of $(N+1)$ coefficients that minimize the maximum deviation from the ideal. Intuitively, this finds the filter that is as close as possible to the desired response given that only $(N+1)$ coefficients can be used. This method is particularly easy in practice since at least one text includes a program that takes the desired filter and N, and returns the optimum coefficients.

- Equiripple FIR filters can be designed using the FFT algorithms as well. The algorithm is iterative in nature. The DFT of an initial filter design is computed using the FFT algorithm (if an initial estimate is not available, h[n]=delta[n] can be used). In the Fourier domain or FFT domain the frequency response is corrected according to the desired specs, and the inverse FFT is then computed. In the time-domain, only the first N coefficients are kept (the other coefficients are set to zero). The process is then repeated iteratively: the FFT is computed once again, correction applied in the frequency domain and so on.

Software packages like MATLAB, GNU Octave, Scilab, and SciPy provide convenient ways to apply these different methods.

Window Design Method

In the window design method, one first designs an ideal IIR filter and then truncates the infinite impulse response by multiplying it with a finite length window function. The result is a finite impulse response filter whose frequency response is modified from that of the IIR filter. Multiplying the infinite impulse by the window function in the time domain results in the frequency response of the IIR being convolved with the Fourier transform (or DTFT) of the window function. If the window's main lobe is narrow, the composite frequency response remains close to that of the ideal IIR filter.

The ideal response is usually rectangular, and the corresponding IIR is a sinc function. The result of the frequency domain convolution is that the edges of the rectangle are tapered, and ripples appear in the passband and stopband. Working backward, one can specify the slope (or width) of the tapered region (*transition band*) and the height of the ripples, and thereby derive the frequency domain parameters of an appropriate window function. Continuing backward to an impulse response can be done by iterating a filter design program to find the minimum filter order. Another method is to restrict the solution set to the parametric family of Kaiser windows, which provides closed form relationships between the time-domain and frequency domain parameters. In general, that method will not achieve the minimum possible filter order, but it is particularly convenient for automated applications that require dynamic, on-the-fly, filter design.

The window design method is also advantageous for creating efficient half-band filters, because the corresponding sinc function is zero at every other sample point (except the center one). The product with the window function does not alter the zeros, so almost half of the coefficients of the final impulse response are zero. An appropriate implementation of the FIR calculations can exploit that property to double the filter's efficiency.

Moving Average Example

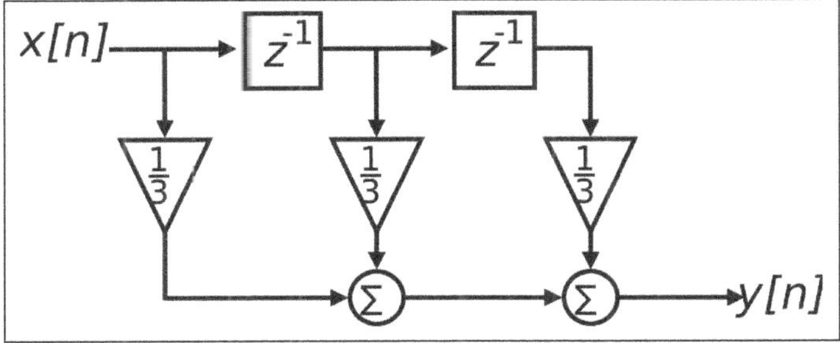

(a) Block diagram of a simple FIR filter (2nd-order/3-tap filter in this case, implementing a moving average).

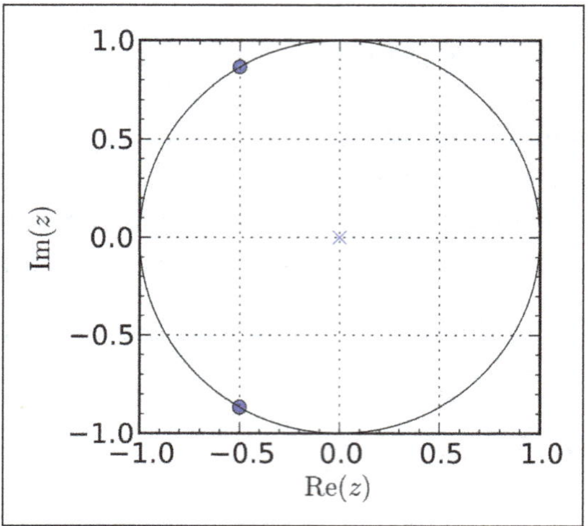

(b) Pole–zero diagram of a second-order FIR filter.

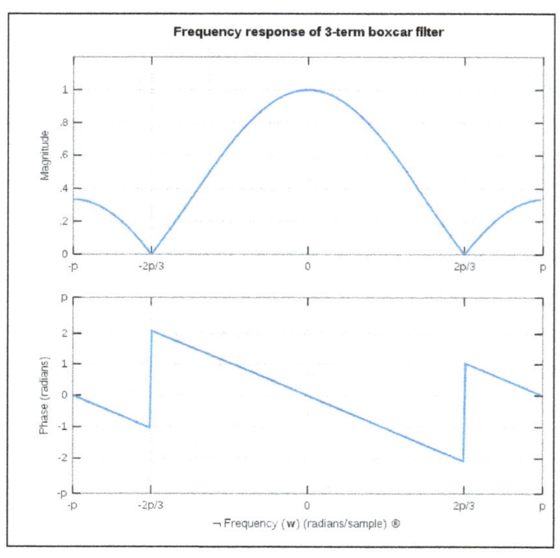

(c) Magnitude and phase responses.

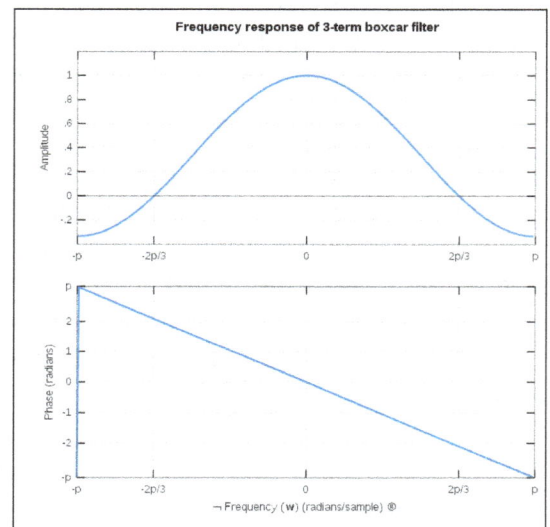

(d) Amplitude and phase responses.

A moving average filter is a very simple FIR filter. It is sometimes called a boxcar filter, especially when followed by decimation. The filter coefficients, $b_0,\ldots,b_N$, are found via the following equation:

$$b_i = \frac{1}{N+1}$$

To provide a more specific example, we select the filter order:

$$N = 2$$

The impulse response of the resulting filter is:

$$h[n] = \frac{1}{3}\delta[n] + \frac{1}{3}\delta[n-1] + \frac{1}{3}\delta[n-2]$$

The fig. (a) on the right shows the block diagram of a 2nd-order moving-average filter. The transfer function is:

$$H(z) = \frac{1}{3} + \frac{1}{3}z^{-1} + \frac{1}{3}z^{-2} = \frac{1}{3}\frac{z^2 + z + 1}{z^2}.$$

fig. (b) on the right shows the corresponding pole–zero diagram. Zero frequency (DC) corresponds to (1, 0), positive frequencies advancing counterclockwise around the circle to the Nyquist frequency at (−1, 0). Two poles are located at the origin, and two zeros are located at $z_1 = -\frac{1}{2} + j\frac{\sqrt{3}}{2}$, $z_2 = -\frac{1}{2} - j\frac{\sqrt{3}}{2}$.

The frequency response, in terms of normalized frequency ω, is:

$$H\left(e^{j\omega}\right) = \frac{1}{3} + \frac{1}{3}e^{-j\omega} + \frac{1}{3}e^{-j2\omega}.$$

fig. (c) on the right shows the magnitude and phase components of $H\left(e^{j\omega}\right)$. But plots like these can also be generated by doing a discrete Fourier transform (DFT) of the impulse response. And because of symmetry, filter design or viewing software often displays only the [0, π] region. The magnitude plot indicates that the moving-average filter passes low frequencies with a gain near 1 and attenuates high frequencies, and is thus a crude low-pass filter. The phase plot is linear except for discontinuities at the two frequencies where the magnitude goes to zero. The size of the discontinuities is π, representing a sign reversal. They do not affect the property of linear phase. That fact is illustrated in fig. (d).

Infinite Impulse Response System

The impulse response, even of IIR systems, usually approaches zero and can be neglected past a certain point. However the physical systems which give rise to IIR or FIR responses are dissimilar, and therein lies the importance of the distinction. For instance, analog electronic filters composed of resistors, capacitors, and/or inductors (and perhaps linear amplifiers) are generally IIR filters. On the other hand, discrete-time filters (usually digital filters) based on a tapped delay line *employing no feedback* are necessarily FIR filters. The capacitors (or inductors) in the analog filter have a "memory" and their internal state never completely relaxes following an impulse (assuming the classical model of capacitors and inductors where quantum effects are ignored). But in the latter case, after an impulse has reached the end of the tapped delay line, the system has no further memory of that impulse and has returned to its initial state; its impulse response beyond that point is exactly zero.

Implementation and Design

Although almost all analog electronic filters are IIR, digital filters may be either IIR or FIR. The presence of feedback in the topology of a discrete-time filter generally creates an IIR response. The z domain transfer function of an IIR filter contains a non-trivial denominator, describing those feedback terms. The transfer function of an FIR filter, on the other hand, has only a numerator as expressed in the general form derived below. All of the a_i coefficients with $i > 0$ (feedback terms) are zero and the filter has no finite poles.

The transfer functions pertaining to IIR analog electronic filters have been extensively studied and optimized for their amplitude and phase characteristics. These continuous-time filter functions are described in the Laplace domain. Desired solutions can be transferred to the case of discrete-time filters whose transfer functions are expressed in the z domain, through the use of certain mathematical techniques such as the bilinear transform, impulse invariance, or pole–zero matching method. Thus digital IIR filters can be based on well-known solutions for analog filters such as the Chebyshev filter, Butterworth filter, and elliptic filter, inheriting the characteristics of those solutions.

Transfer Function Derivation

Digital filters are often described and implemented in terms of the difference equation that defines how the output signal is related to the input signal:

$$\begin{aligned} y[n] = \frac{1}{a_0}&(b_0 x[n] + b_1 x[n-1] + \cdots + b_P x[n-P] \\ &- a_1 y[n-1] - a_2 y[n-2] - \cdots - a_Q y[n-Q]) \end{aligned}$$

where:

- P is the feedforward filter order.
- b_i are the feedforward filter coefficients.
- Q is the feedback filter order.
- a_i are the feedback filter coefficients.
- $x[n]$ is the input signal.
- $y[n]$ is the output signal.

A more condensed form of the difference equation is:

$$y[n] = \frac{1}{a_0}\left(\sum_{i=0}^{P} b_i x[n-i] - \sum_{j=1}^{Q} a_j y[n-j]\right)$$

which, when rearranged, becomes:

$$\sum_{j=0}^{Q} a_j y[n-j] = \sum_{i=0}^{P} b_i x[n-i]$$

To find the transfer function of the filter, we first take the Z-transform of each side of the above equation, where we use the time-shift property to obtain:

$$\sum_{j=0}^{Q} a_j z^{-j} Y(z) = \sum_{i=0}^{P} b_i z^{-i} X(z)$$

We define the transfer function to be:

$$H(z) = \frac{Y(z)}{X(z)}$$

$$= \frac{\sum_{i=0}^{P} b_i z^{-i}}{\sum_{i=0}^{P} a_j z^{-j}}$$

Considering that in most IIR filter designs coefficient a_0 is 1, the IIR filter transfer function takes the more traditional form:

$$H(z) \quad = \frac{\sum_{i=0}^{P} b_i z^{-i}}{1 + \sum_{j=1}^{Q} a_j z^{-j}}$$

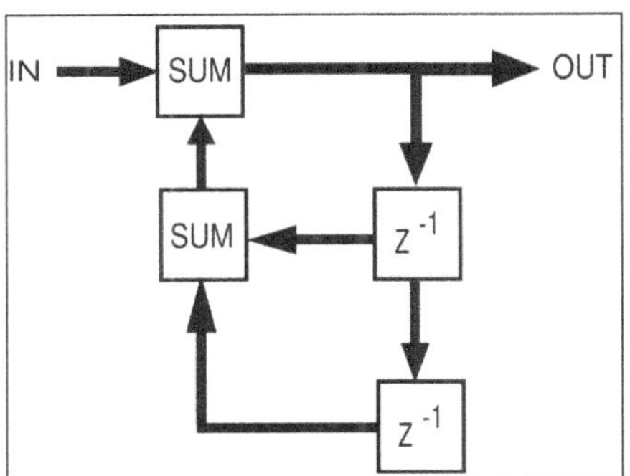

An example of a block diagram of an IIR filter. The z^{-1} block is a unit delay.

Stability

The transfer function allows one to judge whether or not a system is bounded-input, bounded-output (BIBO) stable. To be specific, the BIBO stability criterion requires that the ROC of the system includes the unit circle. For example, for a causal system, all poles of the transfer function have to have an absolute value smaller than one. In other words, all poles must be located within a unit circle in the z -plane.

The poles are defined as the values of z which make the denominator of $H(z)$ equal to 0:

$$0 = \sum_{j=0}^{Q} a_j z^{-j}$$

Clearly, if $a_j \neq 0$ then the poles are not located at the origin of the z -plane. This is in contrast to the FIR filter where all poles are located at the origin, and is therefore always stable.

IIR filters are sometimes preferred over FIR filters because an IIR filter can achieve a much sharper transition region roll-off than an FIR filter of the same order.

Example:

Let the transfer function $H(z)$ of a discrete-time filter be given by:

$$H(z) = \frac{B(z)}{A(z)} = \frac{1}{1 - az^{-1}}$$

governed by the parameter a, a real number with $0 < |a| < 1$. $H(z)$ is stable and causal with a pole at a. The time-domain impulse response can be shown to be given by:

$$h(n) = a^n u(n)$$

where $u(n)$ is the unit step function. It can be seen that $h(n)$ is non-zero for all $n \geq 0$, thus an impulse response which continues infinitely.

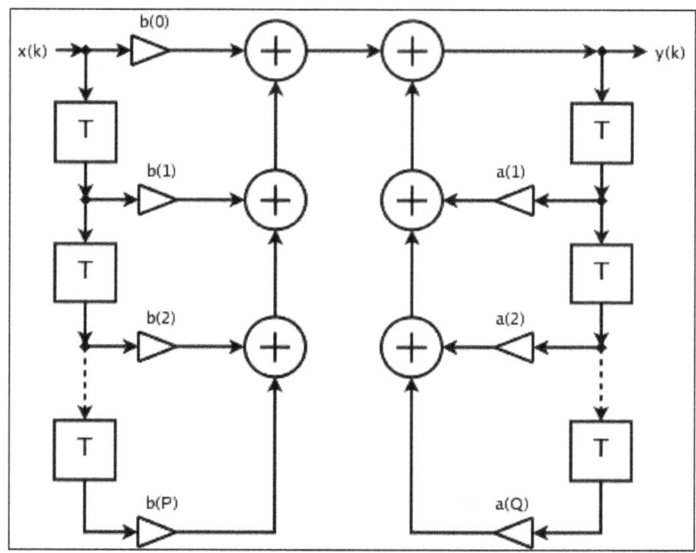

IIR filter example.

Advantages and Disadvantages

The main advantage digital IIR filters have over FIR filters is their efficiency in implementation, in order to meet a specification in terms of passband, stopband, ripple, and/or roll-off. Such a set of specifications can be accomplished with a lower order (Q in the above formulae) IIR filter than would be required for an FIR filter meeting the same requirements. If implemented in a signal processor, this implies a correspondingly fewer number of calculations per time step; the computational savings is often of a rather large factor.

On the other hand, FIR filters can be easier to design, for instance, to match a particular frequency response requirement. This is particularly true when the requirement is not one of the usual cases (high-pass, low-pass, notch, etc.) which have been studied and optimized for analog filters. Also FIR filters can be easily made to be linear phase (constant group delay vs frequency)—a property that is not easily met using IIR filters and then only as an approximation (for instance with the Bessel filter). Another issue regarding digital IIR filters is the potential for limit cycle behavior when idle, due to the feedback system in conjunction with quantization.

IIR and FIR Systems

IIR Systems

- IIR stands for infinite impulse response systems.
- IIR filters are less powerful than FIR filters, & require less processing power and less work to set up the filters.
- They are more easy to change "on the fly".
- These are less flexible.
- It cannot implement linear-phase filtering.
- It cannot be used to correct frequency-response errors in a loudspeaker.
- IIRs can provide good resolution even at low frequencies.
- Usage is generally more easier than FIR filters.

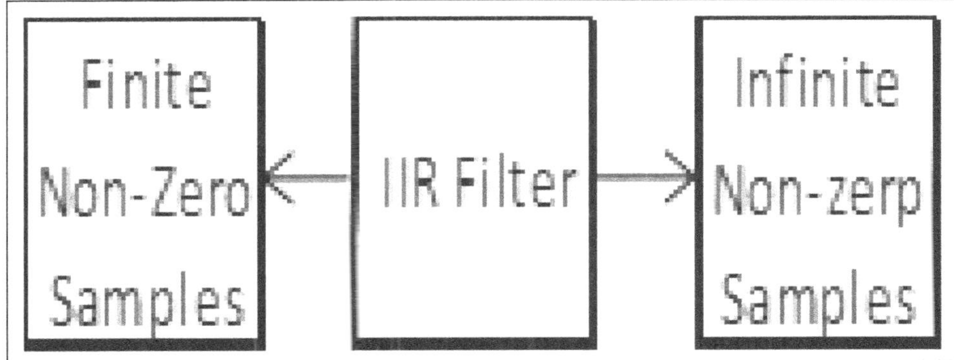

- IIR filter uses current input sample value, past input and output samples to obtain current output sample value.
- Simple IIR equation is mention below. y(n)= b(0)x(n) + b(1)x(n-1) + b(2)x(n-2) + b(3)x(n-3) + a(1)y(n-1) + a(2)y(n-2) + a(3)y(n-3).
- Transfer function of IIR filter will have both zeros and poles and will require less memory than FIR counterpart.
- IIR filters are not stable as they are recursive in nature and feedback is also involved in the process of calculating output sample values.
- IIR filter need more power due to more coefficients in the design.
- IIR filters have analog equivalent.

- IIR filters are more efficient.
- IIR filters are used as notch(band stop),band pass functions.
- IIR filter need lower order than FIR filter to achieve same performance.
- Delay is less than FIR filter.
- It has higher sensitivity than FIR filter.

FIR Systems

- FIR stands for finite impulse response systems.
- FIR filters are more powerful than IIR filters, but also require more processing power and more work to set up the filters.
- They are also less easy to change "on the fly" as you can by tweaking (say) the frequency setting of a parametric (IIR) filter.
- Their, greater power means more flexibility and ability to finely adjust the response of your active loudspeaker.
- It can implement linear-phase filtering.
- It can be used to correct frequency-response errors in a loudspeaker to a finer degree of precision than using IIRs.
- FIRs can be limited in resolution at low frequencies, and the success of applying FIR filters depends greatly on the program that is used to generate the filter coefficients.
- Usage is generally more complicated and time-consuming than IIR filters.

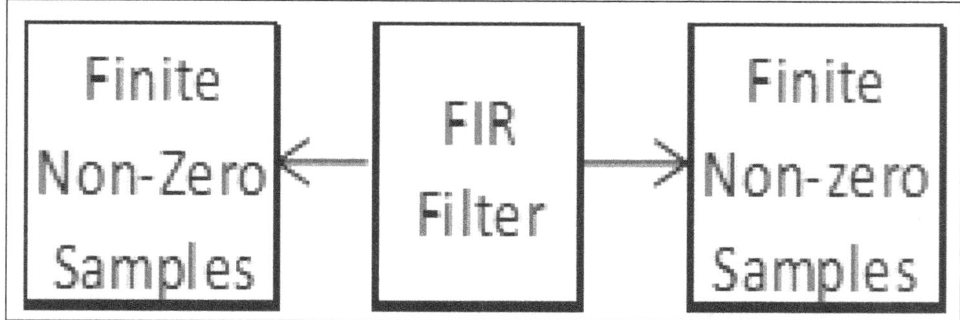

- FIR filter uses only current and past input digital samples to obtain a current output sample value. It does not utilize past output samples.
- Simple FIR equation is: $y(n)= h(0)x(n) + h(1)x(n-1) + h(2)x(n-2) + h(3)x(n-3) + h(4)x(n-4)$.
- Transfer function of FIR filter will have only zeros, need more memory.

- FIR filters are preferred due to its linear phase response and also they are non-recursive. Feedback is not involved in FIR, hence they are stable.
- FIR filter consume low power.
- FIR have no analog equivalent.
- FIR filters are less efficient.
- FIR filters are used as anti-aliasing, low pass and baseband filters.
- FIR filter need higher order than IIR filter to achieve same performance.
- Delay is more than IIR filter.
- It has lower sensitivity than IIR filter.

Time Variant and Time Invariant Systems

A system is said to be time variant if its input and output characteristics vary with time. Otherwise, the system is considered as time invariant.

The condition for time invariant system is:

$$y(n,t) = y(n-t)$$

The condition for time variant system is:

$$y(n,t) \neq y(n-t)$$

Where,

$$y(n,t) = T\left[x(n-t)\right] = \text{input change}$$

$$y(n-t) = \text{output change}$$

Example:

$$y(n) = x(-n)$$

$$y(n,\ t) = T\left[x(n-t)\right] = x(-n-t)$$

$$y(n,\ t) = T\left[x(n-t)\right] = x(-n-t)$$

$$\therefore y(n,\ t) \neq y(n-t).$$

Hence, the system is time variant.

References

- Classification-of-systems, signals-systems: ecetutorials.com, Retrieved 16 March, 2019
- Static-dynamic-systems: myclassbook.org, Retrieved 17 April, 2019
- Systems-classification, signals-and-systems: tutorialspoint.com, Retrieved 13 August, 2019
- Differentiate-iir-and-fir-systems- 13648: ques10.com, Retrieved 14 July, 2019
- Linear-non-linear-systems-linearity-property: myclassbook.org, Retrieved 18 June, 2019
- Classification-of-systems, signals-systems: ecetutorials.com, Retrieved 15 February, 2019

4

Fourier Series and Fourier Transform

Fourier series represents an expansion of periodic operation in terms of an infinite sum sines and cosines. Fourier transform converts a general and non-periodic operation into its constituent frequencies. The topics elaborated in this chapter will help in gaining a better perspective about the fourier series and fourier transform.

To represent any periodic signal x(t), Fourier developed an expression called Fourier series. This is in terms of an infinite sum of sines and cosines or exponentials. Fourier series uses orthoganality condition.

Fourier Series Representation of Continuous Time Periodic Signals

A signal is said to be periodic if it satisfies the condition x (t) = x (t + T) or x (n) = x (n + N).

Where T = fundamental time period.

ω_o = fundamental frequency = 2π/T.

There are two basic periodic signals.

1. $x(t) = \cos\ \omega_0 t$ (sinusoidal).
2. $x(t) = e^{j\omega 0 t}$ (complex exponential).

These two signals are periodic with period:

$$T = 2\pi / \omega_0$$

A set of harmonically related complex exponentials can be represented as $\{\phi_k(t)\}$.

$$\phi_k(t) = \{e^{jk\omega_0 t}\} = \{e^{jk(\frac{2\pi}{T})t}\} \text{where } k = 0 \pm 1, \pm 2..n$$

All these signals are periodic with period T.

According to orthogonal signal space approximation of a function x (t) with n, mutually orthogonal functions is given by:

$$x(t) = \sum_{k=-\infty}^{\infty} a_k e^{jk\omega_0 t}$$

$$= \sum_{k=-\infty}^{\infty} a_k k e^{jk\omega_0 t}$$

Where a_k = Fourier coefficient = coefficient of approximation.

This signal $x(t)$ is also periodic with period T.

Equation 2 represents Fourier series representation of periodic signal x(t).

The term k = 0 is constant.

The term $k = \pm 1$ having fundamental frequency ω_0, is called as 1st harmonics.

The term $k = \pm 2$ having fundamental frequency $2\omega_0$, is called as 2nd harmonics, and so on.

The term $k = \pm n$ having fundamental frequency $n\omega 0$, is called as nth harmonics.

Deriving Fourier Coefficient

We know that $x(t) = \sum_{k=-\infty}^{\infty} a_k e^{jk\omega_0 t}$

Multiply $e^{-jn\omega_0 t}$ on both sides.

$$x(t)e^{-jn\omega_0 t} = \sum_{k=-\infty}^{\infty} a_k e^{jk\omega_0 t}.e^{-jn\omega_0 t}$$

Consider integral on both sides.

$$\int_0^T x(t)e^{jk\omega_0 t}dt = \int_0^T \sum_{k=-\infty}^{\infty} a_k e^{jk\omega_0 t}.e^{-jn\omega_0 t}dt$$

$$= \int_0^T \sum_{k=-\infty}^{\infty} a_k e^{j(k-n)\omega_0 t}.dt$$

$$\int_0^T x(t)e^{jk\omega_0 t}dt = \sum_{k=-\infty}^{\infty} a_k \int_0^T e^{j(k-n)\omega_0 t}.dt$$

by Euler's formula:

$$\int_0^T e^{j(k-n)\omega_0 t}dt. = \int_0^T \cos(k-n)\omega_0 dt + j\int_0^T \sin(k-n)\omega_0 t\, dt$$

$$\int_0^T e^{j(k-n)\omega_0 t} dt. = \begin{cases} T & k=n \\ 0 & k \neq n \end{cases}$$

Hence in equation, the integral is zero for all values of k except at k = n. Put k = n in equation:

$$\Rightarrow \int_0^T x(t) e^{-jn\omega_0 t} dt = a_n T$$

$$\Rightarrow a_n = \frac{1}{T} \int_0^T e^{-jn\omega_0 t} dt$$

Replace n by k.

$$\Rightarrow a_k = \frac{1}{T} \int_0^T e^{-jk\omega_0 t} dt$$

$$\therefore x(t) = \sum_{k=-\infty}^{\infty} a_k e^{j(k-n)\omega_0 t}$$

where $a_k = \frac{1}{T} \int_0^T e^{-jk\omega_0 t} dt$

Properties of Fourier Series

These are properties of Fourier series:

Linearity Property

If $x(t) \xleftrightarrow{\text{fourier series coefficient}} f_{xn}$ and $y(t) \xleftrightarrow{\text{fourier series coefficient}} f_{yn}$

then linearity property states that,

$$\text{a } x(t) + \text{b } y(t) \xleftrightarrow{\text{fourier series coefficient}} \text{a } fxn + \text{b } fyn$$

Time Shifting Property

If $x(t) \xleftrightarrow{\text{fourier series coefficient}} f_{xn}$

then time shifting property states that:

$$x(t - t_0) \xleftrightarrow{\text{fourier series coefficient}} e^{-jn\omega_0 t_0} f_{xn}$$

Frequency Shifting Property

If $x(t) \xleftrightarrow{\text{fourier series coefficient}} f_{xn}$

then frequency shifting property states that:

$$e^{-jn\omega_0 t_0} . x(t) \xleftrightarrow{\text{fourier series coefficient}} f_{x(n-n_0)}$$

Time Reversal Property

If $x(t) \xleftrightarrow{\text{fourier series coefficient}} f_{xn}$.

Then time reversal property states that:

$$x(-t) \xleftrightarrow{\text{fourier series coefficient}} f_{-xn}$$

Time Scaling Property

If $x(t) \xleftrightarrow{\text{fourier series coefficient}} f_{xn}$

then time Scaling property states that:

If $x(at) \xleftrightarrow{\text{fourier series coefficient}} f_{xn}$

Time scaling property changes frequency components from ω_0 to $a\omega_0$.

Differentiation and Integration Properties

If $x(t) \xleftrightarrow{\text{fourier series coefficient}} f_{xn}$

then differentiation property states that:

If $\frac{dx(t)}{dt} \xleftrightarrow{\text{fourier series coefficient}} jn\omega_0.f_{xn}$

& integration property states that:

If $\int x(t)dt \xleftrightarrow{\text{fourier series coefficient}} jn\omega_0.f_{xn}$

Multiplication and Convolution Properties

If $x(t) \xleftrightarrow{\text{fourier series coefficient}} f_{xn}$ and $y(t) \xleftrightarrow{\text{fourier series coefficient}} f_{yn}$

Then multiplication property states that:

$$x(t).y(t) \xleftrightarrow{\text{fourier series coefficient}} T\, f_{xn} * f_{yn}$$

Convolution property states that:

$$x(t) * y(t) \xleftrightarrow{\text{fourier series coefficient}} T\, f_{xn}.f_{yn}$$

Conjugate and Conjugate Symmetry Properties

If $x(t) \xleftrightarrow{\text{fourier series coefficient}} f_{xn}$

Then conjugate property states that:

$$x*(t) \xleftrightarrow{\text{fourier series coefficient}} f*_{xn}$$

Conjugate symmetry property for real valued time signal states that:

$$f*_{xn} = f - xn$$

& Conjugate symmetry property for imaginary valued time signal states that:

$$f*_{xn} = -f_{-xn}$$

Fourier Transform

The Fourier transform (FT) decomposes a function of time (a *signal*) into its constituent frequencies. This is similar to the way a musical chord can be expressed in terms of the volumes and frequencies of its constituent notes. The term *Fourier transform* refers to both the frequency domain representation and the mathematical operation that associates the frequency domain representation to a function of time. The Fourier transform of a function of time is itself a complex-valued function of frequency, whose magnitude (modulus) represents the amount of that frequency present in the original function, and whose argument is the phase offset of the basic sinusoid in that frequency. The Fourier transform is not limited to functions of time, but the domain of the original function is commonly referred to as the *time domain*. There is also an *inverse Fourier transform* that mathematically synthesizes the original function from its frequency domain representation.

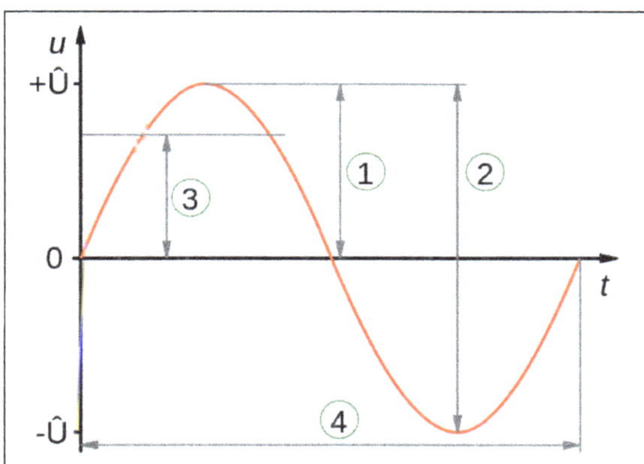

A sinusoidal curve, with peak amplitude (1), peak-to-peak (2), RMS (3), and wave period (4).

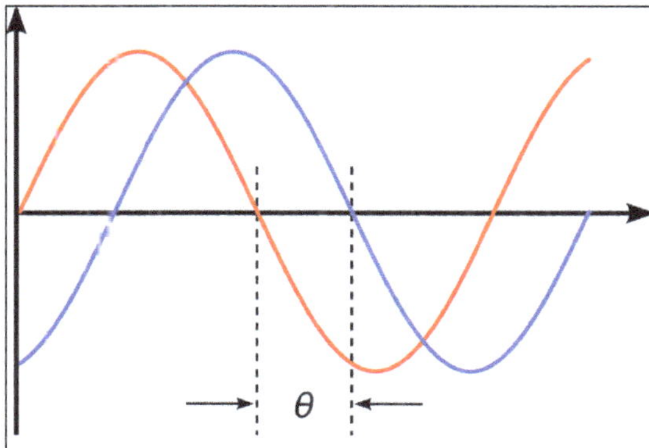

Illustration of phase shift θ.

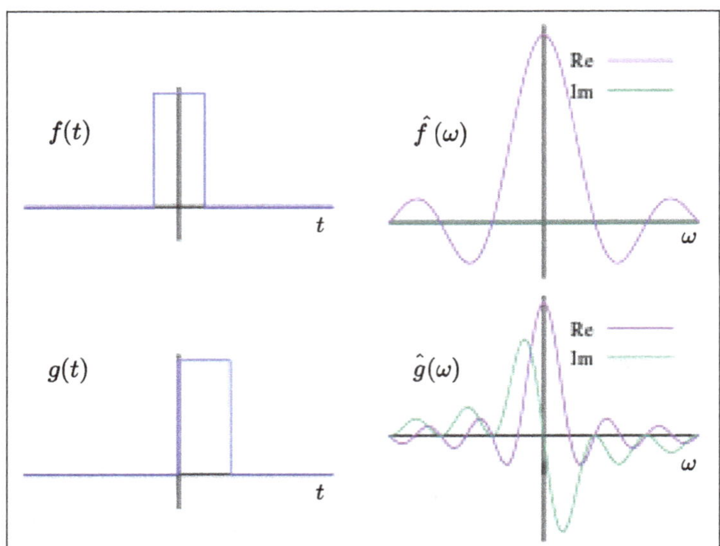

In the first row of the figure is the graph of the unit pulse function $f(t)$ and its Fourier transform $\hat{f}(\omega)$, a function of frequency ω. Translation (that is, delay) in the time domain is interpreted as complex phase shifts in the frequency domain. In the second row is shown $g(t)$, a delayed unit pulse, beside the real and imaginary parts of the Fourier transform. The Fourier transform decomposes a function into eigenfunctions for the group of translations.

Linear operations performed in one domain (time or frequency) have corresponding operations in the other domain, which are sometimes easier to perform. The operation of differentiation in the time domain corresponds to multiplication by the frequency, so some differential equations are easier to analyze in the frequency domain. Also, convolution in the time domain corresponds to ordinary multiplication in the frequency domain. After performing the desired operations, transformation of the result can be made back to the time domain. Harmonic analysis is the systematic study of the relationship between the frequency and time domains, including the kinds of functions or operations that are "simpler" in one or the other, and has deep connections to many areas of modern mathematics.

Functions that are localized in the time domain have Fourier transforms that are spread out across the frequency domain and vice versa, a phenomenon known as the uncertainty principle. The critical case for this principle is the Gaussian function, of substantial importance in probability theory and statistics as well as in the study of physical phenomena exhibiting normal distribution (e.g., diffusion). The Fourier transform of a Gaussian function is another Gaussian function. Joseph Fourier introduced the transform in his study of heat transfer, where Gaussian functions appear as solutions of the heat equation.

The Fourier transform can be formally defined as an improper Riemann integral, making it an integral transform, although this definition is not suitable for many applications requiring a more sophisticated integration theory. For example, many relatively simple applications use the Dirac delta function, which can be treated formally as if it were a function, but the justification requires a mathematically more sophisticated viewpoint. The Fourier transform can also be generalized to functions of several variables on Euclidean space, sending a function of 3-dimensional 'position space' to a function of 3-dimensional momentum (or a function of space and time to a function of 4-momentum). This idea makes the spatial Fourier transform very natural in the study of waves, as well as in quantum mechanics, where it is important to be able to represent wave solutions as functions of either position or momentum and sometimes both. In general, functions to which Fourier methods are applicable are complex-valued, and possibly vector-valued. Still further generalization is possible to functions on groups, which, besides the original Fourier transform on R or R^n (viewed as groups under addition), notably includes the discrete-time Fourier transform (DTFT, group = Z), the discrete Fourier transform (DFT, group = Z mod N) and the Fourier series or circular Fourier transform (group = S^1, the unit circle ≈ closed finite interval with endpoints identified). The latter is routinely employed to handle periodic functions. The fast Fourier transform (FFT) is an algorithm for computing the DFT.

The Fourier transform of a function f is traditionally denoted $\hat{f}$, by adding a circumflex to the symbol of the function. There are several common conventions for defining the Fourier transform of an integrable function $f:\mathbb{R}\to\mathbb{C}$. One of them is:

$$\hat{f}(\xi)=\int_{-\infty}^{\infty} f(x)\, e^{-2\pi i x\xi}\, dx,$$

for any real number ξ.

A reason for the negative sign in the exponent is that it is common in electrical engineering to represent by $f(x)=e^{2\pi i\xi_0 x}$ a signal with zero initial phase and frequency ξ_0. The negative sign convention causes the product $e^{2\pi i\xi_0 x}e^{-2\pi i\xi x}$ to be 1 (frequency zero) when $\xi=\xi_0$, causing the integral to diverge. The result is a Dirac delta function at $\xi=\xi_0$, which is the only frequency component of the sinusoidal signal $e^{2\pi i\xi_0 x}$.

When the independent variable x represents *time*, the transform variable ξ represents frequency (e.g. if time is measured in seconds then frequency is in hertz). Under suitable conditions, f is determined by $\hat{f}$ via the inverse transform:

$$f(x)=\int_{-\infty}^{\infty} \hat{f}(\xi)\, e^{2\pi i x\xi}\, d\xi,$$

for any real number x.

The statement that f can be reconstructed from $\hat{f}$ is known as the Fourier inversion theorem, and was first introduced in Fourier's Analytical Theory of Heat, although what would be considered a proof by modern standards was not given until much later. The functions f and $\hat{f}$ often are referred to as a Fourier integral pair or Fourier transform pair.

For other common conventions and notations, including using the angular frequency ω instead of the frequency ξ, The Fourier transform on Euclidean space is treated separately, in which the variable x often represents position and ξ momentum. The conventions chosen here are those of harmonic analysis, and are characterized as the unique conventions such that the Fourier transform is both unitary on L^2 and an algebra homomorphism from L^1 to L^∞, without renormalizing the Lebesgue measure.

Many other characterizations of the Fourier transform exist. For example, one uses the Stone–von

Neumann theorem: the Fourier transform is the unique unitary intertwiner for the symplectic and Euclidean Schrödinger representations of the Heisenberg group.

One motivation for the Fourier transform comes from the study of Fourier series. In the study of Fourier series, complicated but periodic functions are written as the sum of simple waves mathematically represented by sines and cosines. The Fourier transform is an extension of the Fourier series that results when the period of the represented function is lengthened and allowed to approach infinity.

Due to the properties of sine and cosine, it is possible to recover the amplitude of each wave in a Fourier series using an integral. In many cases it is desirable to use Euler's formula, which states that $e^{2\pi i\theta} = \cos(2\pi\theta) + i\sin(2\pi\theta)$, to write Fourier series in terms of the basic waves $e^{2\pi i\theta}$. This has the advantage of simplifying many of the formulas involved, and provides a formulation for Fourier series that more closely resembles the definition followed here. Re-writing sines and cosines as complex exponentials makes it necessary for the Fourier coefficients to be complex valued. The usual interpretation of this complex number is that it gives both the amplitude (or size) of the wave present in the function and the phase (or the initial angle) of the wave. These complex exponentials sometimes contain negative "frequencies". If θ is measured in seconds, then the waves $e^{2\pi i\theta}$ and $e^{-2\pi i\theta}$ both complete one cycle per second, but they represent different frequencies in the Fourier transform. Hence, frequency no longer measures the number of cycles per unit time, but is still closely related.

There is a close connection between the definition of Fourier series and the Fourier transform for functions f that are zero outside an interval. For such a function, we can calculate its Fourier series on any interval that includes the points where f is not identically zero. The Fourier transform is also defined for such a function. As we increase the length of the interval in which we calculate the Fourier series, then the Fourier series coefficients begin to resemble the Fourier transform and the sum of the Fourier series of f begins to resemble the inverse Fourier transform. More precisely, suppose T is large enough that the interval $[-T/2, T/2]$ contains the interval in which f is not identically zero. Then, the nth series coefficient c_n is given by:

$$c_n = \frac{1}{T}\int_{-\frac{T}{2}}^{\frac{T}{2}} f(x)e^{-2\pi i\left(\frac{n}{T}\right)x}\,dx.$$

Comparing this to the definition of the Fourier transform, it follows that:

$$c_n = \frac{1}{T}\hat{f}\left(\frac{n}{T}\right)$$

since $f(x)$ is zero outside $[-\frac{T}{2}, \frac{T}{2}]$. Thus, the Fourier coefficients are equal to the values of the Fourier transform sampled on a grid of width $\frac{1}{T}$, multiplied by the grid width $\frac{1}{T}$.

Under appropriate conditions, the Fourier series of f will equal the function f. In other words, f can be written:

$$f(x) = \sum_{n=-\infty}^{\infty} c_n e^{2\pi i\left(\frac{n}{T}\right)x} = \sum_{n=-\infty}^{\infty} \hat{f}(\xi_n)\, e^{2\pi i\xi_n x}\Delta\xi,$$

where the last sum is simply the first sum rewritten using the definitions $\xi_n = \frac{n}{T}$, and $\Delta\xi = \frac{n+1}{T} - \frac{n}{T} = \frac{1}{T}$.

This second sum is a Riemann sum. By letting $T \to \infty$ it will converge to the integral for the inverse Fourier transform as expressed above. Under suitable conditions, this argument may be made precise.

In the study of Fourier series the numbers c_n could be thought of as the "amount" of the wave present in the Fourier series of f. Similarly, the Fourier transform can be thought of as a function that measures how much of each individual frequency is present in our function f, and we can recombine these waves by using an integral (or "continuous sum") to reproduce the original function.

Example:

The following figures provide a visual illustration how the Fourier transform measures whether a frequency is present in a particular function. The depicted function $f(t) = \cos(6\pi t)\, e^{-\pi t2}$ oscillates at 3 Hz (if t measures seconds) and tends quickly to 0. (The second factor in this equation is an envelope function that shapes the continuous sinusoid into a short pulse. Its general form is a Gaussian function). This function was specially chosen to have a real Fourier transform that can be easily plotted. The first image contains its graph. In order to calculate $\hat{f}$ we must integrate $e^{-2\pi i(3t)}$ $f(t)$. The second image shows the plot of the real and imaginary parts of this function. The real part of the integrand is almost always positive, because when $f(t)$ is negative, the real part of $e^{-2\pi i(3t)}$ is negative as well. Because they oscillate at the same rate, when $f(t)$ is positive, so is the real part of $e^{-2\pi i(3t)}$. The result is that when you integrate the real part of the integrand you get a relatively large number (in this case $\frac{1}{2}$). On the other hand, when you try to measure a frequency that is not present, as in the case when we look at $\hat{f}$, you see that both real and imaginary component of this function vary rapidly between positive and negative values, as plotted in the third image. Therefore, in this case, the integrand oscillates fast enough so that the integral is very small and the value for the Fourier transform for that frequency is nearly zero.

The general situation may be a bit more complicated than this, but this in spirit is how the Fourier transform measures how much of an individual frequency is present in a function $f(t)$.

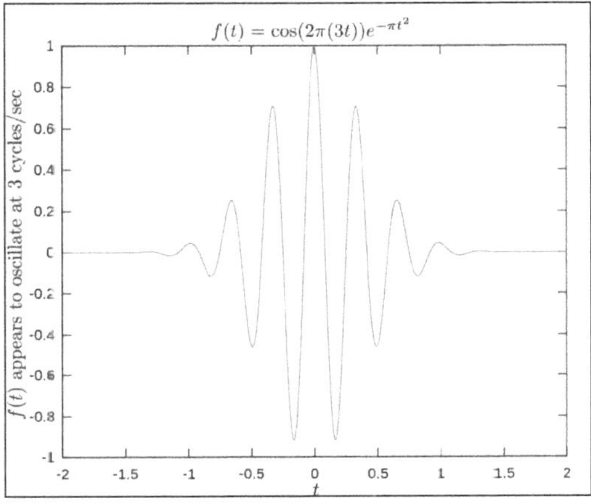

Original function showing oscillation 3 Hz.

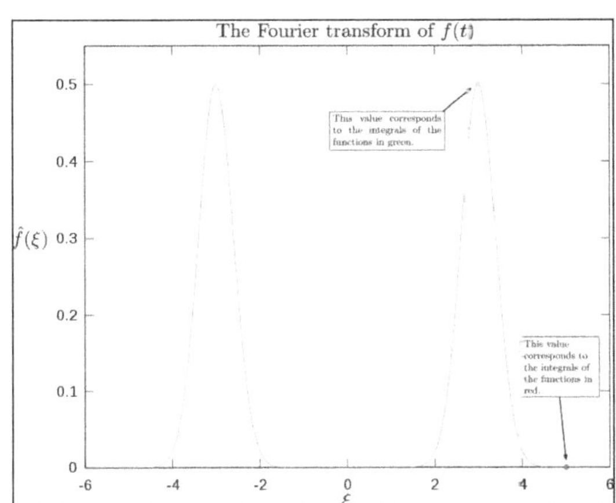

Magnitude of Fourier transform, with 3 and 5 Hz labeled.

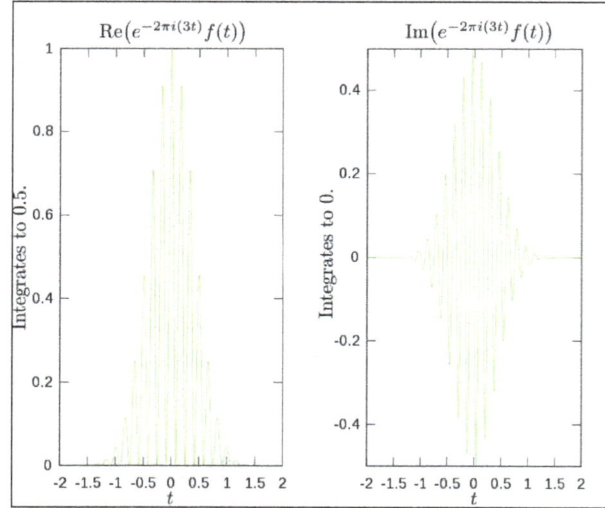

Real and imaginary parts of integrand for Fourier transform at 3 Hz.

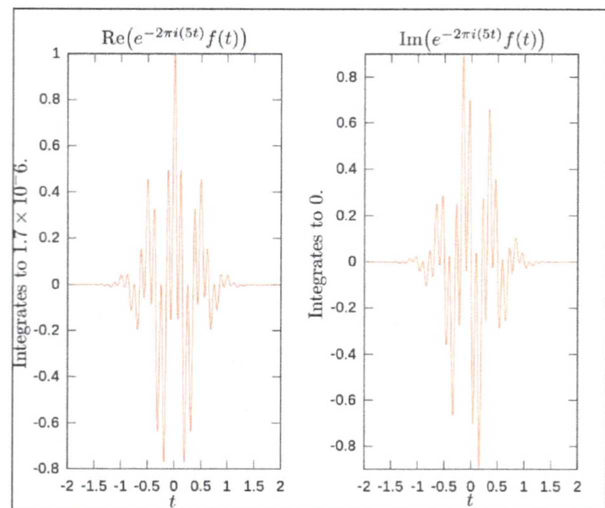

Real and imaginary parts of integrand for Fourier transform at 5 Hz.

Properties of the Fourier Transform

Here we assume $f(x)$, $g(x)$ and $h(x)$ are *integrable functions.*

Lebesgue-measurable on the real line satisfying:

$$\int_{-\infty}^{\infty} |f(x)|dx < \infty.$$

We denote the Fourier transforms of these functions as $\hat{f}(\xi)$, $\hat{g}(\xi)$ and $\hat{h}(\xi)$ respectively.

The Fourier transform has the following basic properties:

Linearity

For any complex numbers a and b, if $h(x) = af(x) + bg(x)$, then $\hat{h}(\xi) = a \cdot \hat{f}(\xi) + b \cdot \hat{g}(\xi)$.

Translation or Time Shifting

For any real number x_0, if $h(x) = f(x - x_0)$, then $\hat{h}(\xi) = e^{-2\pi i x_0 \xi}\hat{f}(\xi)$.

Modulation or Frequency Shifting

For any real number ξ_0, if $h(x) = e^{2\pi i x \xi_0} f(x)$, then $\hat{h}(\xi) = \hat{f}(\xi - \xi_0)$.

Time Scaling

For a non-zero real number a, if $h(x) = f(ax)$.

The case $a = -1$ leads to the time-reversal property, which states: if $h(x) = f(-x)$, then $\hat{h}(\xi) = \hat{f}(-\xi)$.

Conjugation

If h(x) = f(x),

In particular, if f is real, then one has the reality condition.

That is, $\hat{f}$ is a Hermitian function. And if f is purely imaginary.

Real and Imaginary Part in Time

- If $h(x) = \Re(f(x))$, then $\hat{h}(\xi) = \frac{1}{2}(\hat{f}(\xi) + \overline{\hat{f}(-\xi)})$.
- If $h(x) = \Im(f(x))$, then $\hat{h}(\xi) = \frac{1}{2i}(\hat{f}(\xi) - \overline{\hat{f}(-\xi)})$.

Integration

Substituting ξ = 0 in the definition, we obtain:

$$\hat{f}(0) = \int_{-\infty}^{\infty} f(x)dx.$$

That is, the evaluation of the Fourier transform at the origin (ξ = 0) equals the integral of f over all its domain.

Invertibility and Periodicity

Under suitable conditions on the function *f*, it can be recovered from its Fourier transform $\hat{f}$. Indeed, denoting the Fourier transform operator by *F*, so $F(f) := \hat{f}$, then for suitable functions, applying the Fourier transform twice simply flips the function: $F^2(f)(x) = f(-x)$, which can be interpreted as "reversing time". Since reversing time is two-periodic, applying this twice yields $F^4(f) = f$, so the Fourier transform operator is four-periodic, and similarly the inverse Fourier transform can be obtained by applying the Fourier transform three times: $F^3(\hat{f}) = f$. In particular the Fourier transform is invertible (under suitable conditions).

More precisely, defining the *parity operator P* that inverts time, $P[f] : t \mapsto f(-t)$:

$$\mathcal{F}^0 = \mathrm{Id}, \quad \mathcal{F}^1 = \mathcal{F},$$
$$\mathcal{F}^2 = \mathcal{P}, \quad \mathcal{F}^3 = \mathcal{F}^{-1} = \mathcal{P}\circ\mathcal{F} = \mathcal{F}\circ\mathcal{P},$$
$$\mathcal{F}^4 = \mathrm{Id}$$

These equalities of operators require careful definition of the space of functions in question, defining equality of functions (equality at every point? equality almost everywhere?) and defining equality of operators – that is, defining the topology on the function space and operator space in question. These are not true for all functions, but are true under various conditions, which are the content of the various forms of the Fourier inversion theorem.

This fourfold periodicity of the Fourier transform is similar to a rotation of the plane by 90°, particularly as the two-fold iteration yields a reversal, and in fact this analogy can be made precise.

While the Fourier transform can simply be interpreted as switching the time domain and the frequency domain, with the inverse Fourier transform switching them back, more geometrically it can be interpreted as a rotation by 90° in the time–frequency domain (considering time as the x-axis and frequency as the y-axis), and the Fourier transform can be generalized to the fractional Fourier transform, which involves rotations by other angles. This can be further generalized to linear canonical transformations, which can be visualized as the action of the special linear group $SL_2(R)$ on the time–frequency plane, with the preserved symplectic form corresponding to the uncertainty principle, below. This approach is particularly studied in signal processing, under time–frequency analysis.

Units and Duality

In mathematics, one often does not think of any units as being attached to the two variables t and ξ. But in physical applications, ξ must have inverse units to the units of t. For example, if t is measured in seconds, ξ should be in cycles per second for the formulas here to be valid. If the scale of t is changed and t is measured in units of 2π seconds, then either ξ must be in the so-called "angular frequency", or one must insert some constant scale factor into some of the formulas. If t is measured in units of length, then ξ must be in inverse length, e.g., wavenumbers. That is to say, there are two copies of the real line: one measured in one set of units, where t ranges, and the other in inverse units to the units of t, and which is the range of ξ. So these are two distinct copies of the real line, and cannot be identified with each other. Therefore, the Fourier transform goes from one space of functions to a different space of functions: functions which have a different domain of definition.

In general, ξ must always be taken to be a linear form on the space of ts, which is to say that the second real line is the dual space of the first real line. This point of view becomes essential in generalisations of the Fourier transform to general symmetry groups, including the case of Fourier series.

That there is no one preferred way (often, one says "no canonical way") to compare the two copies of the real line which are involved in the Fourier transform—fixing the units on one line does not force the scale of the units on the other line—is the reason for the plethora of rival conventions on the definition of the Fourier transform. The various definitions resulting from different choices of units differ by various constants. If the units of t are in seconds but the units of ξ are in angular frequency, then the angular frequency variable is often denoted by one or another Greek letter, for example, $\omega = 2\pi\xi$ is quite common.

$$\hat{x}_1(\omega) = \hat{x}\left(\frac{\omega}{2\pi}\right) = \int_{-\infty}^{\infty} x(t)e^{-i\omega t}\,dt$$

as before, but the corresponding alternative inversion formula would then have to be:

$$x(t) = \frac{1}{2\pi}\int_{-\infty}^{\infty} \hat{x}_1(\omega)e^{it\omega}\,d\omega.$$

To have something involving angular frequency but with greater symmetry between the Fourier transform and the inversion formula, one very often sees still another alternative definition of the Fourier transform, with a factor of $\sqrt{2\pi}$, thus:

$$\hat{x}_2(\omega) = \frac{1}{\sqrt{2\pi}} \int_{-\infty}^{\infty} x(t) e^{-it\omega} \, dt,$$

and the corresponding inversion formula then has to be:

$$x(t) = \frac{1}{\sqrt{2\pi}} \int_{-\infty}^{\infty} \hat{x}_2(\omega) e^{it\omega} \, d\omega.$$

In some unusual conventions, such as those employed by the FourierTransform command of the Wolfram Language, the Fourier transform has i in the exponent instead of $-i$, and vice versa for the inversion formula. Many of the identities involving the Fourier transform remain valid in those conventions, provided all terms that explicitly involve i have it replaced by $-i$.

For example, in probability theory, the characteristic function ϕ of the probability density function f of a random variable X of continuous type is defined without a negative sign in the exponential, and since the units of x are ignored, there is no 2π either:

$$\phi(\lambda) = \int_{-\infty}^{\infty} f(x) e^{i\lambda x} \, dx.$$

(In probability theory, and in mathematical statistics, the use of the Fourier—Stieltjes transform is preferred, because so many random variables are not of continuous type, and do not possess a density function, and one must treat not functions but distributions, i.e., measures which possess "atoms".)

From the higher point of view of group characters, which is much more abstract, all these arbitrary choices disappear, which treats the notion of the Fourier transform of a function on a locally compact Abelian group.

Uniform continuity and the Riemann–Lebesgue lemma

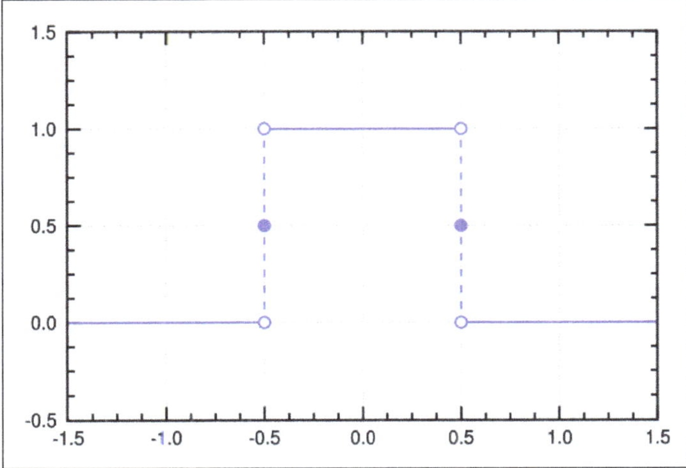

The rectangular function is Lebesgue integrable.

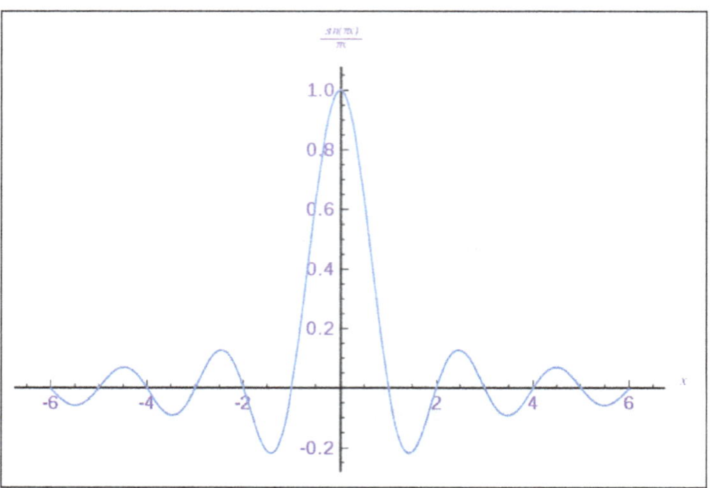

The sinc function, which is the Fourier transform of the rectangular function, is bounded and continuous, but not Lebesgue integrable.

The Fourier transform may be defined in some cases for non-integrable functions, but the Fourier transforms of integrable functions have several strong properties.

The Fourier transform $\hat{f}$ of any integrable function f is uniformly continuous:

$$\left\|\hat{f}\right\|_{\infty} \leq \|f\|_1$$

By the Riemann–Lebesgue lemma:

$$\hat{f}(\xi) \to 0 \text{ as } |\xi| \to \infty.$$

However, $\hat{f}$ need not be integrable. For example, the Fourier transform of the rectangular function, which is integrable, is the sinc function, which is not Lebesgue integrable, because its improper integrals behave analogously to the alternating harmonic series, in converging to a sum without being absolutely convergent.

It is not generally possible to write the inverse transform as a Lebesgue integral. However, when both f and $\hat{f}$ are integrable, the inverse equality:

$$f(x) = \int_{-\infty}^{\infty} \hat{f}(\xi) e^{2i\pi x\xi} \, d\xi$$

holds almost everywhere. That is, the Fourier transform is injective on $L^1(R)$. (But if f is continuous, then equality holds for every x).

Plancherel Theorem and Parseval's Theorem

Let $f(x)$ and $g(x)$ be integrable, and let $\hat{f}(\xi)$ and $\hat{g}(\xi)$ be their Fourier transforms. If $f(x)$ and $g(x)$ are also square-integrable, then the Parseval formula follows:

$$\int_{-\infty}^{\infty} f(x)\overline{g(x)}dx = \int_{-\infty}^{\infty} \hat{f}(\xi)\overline{\hat{g}(\xi)}d\xi,$$

where the bar denotes complex conjugation.

The Plancherel theorem, which follows from the above, states that:

$$\int_{-\infty}^{\infty} |f(x)|^2 dx = \int_{-\infty}^{\infty} |\hat{f}(\xi)|^2 d\xi,$$

Plancherel's theorem makes it possible to extend the Fourier transform, by a continuity argument, to a unitary operator on $L^2(\mathbb{R})$. On $L^1(\mathbb{R}) \cap L^2(\mathbb{R})$, this extension agrees with original Fourier transform defined on $L^1(\mathbb{R})$, thus enlarging the domain of the Fourier transform to $L^1(\mathbb{R}) + L^2(\mathbb{R})$ (and consequently to $L^p(\mathbb{R})$ for $1 \le p \le 2$). Plancherel's theorem has the interpretation in the sciences that the Fourier transform preserves the energy of the original quantity. The terminology of these formulas is not quite standardised. Parseval's theorem was proved only for Fourier series, and was first proved by Lyapunov. But Parseval's formula makes sense for the Fourier transform as well, and so even though in the context of the Fourier transform it was proved by Plancherel, it is still often referred to as Parseval's formula, or Parseval's relation, or even Parseval's theorem.

Poisson Summation Formula

The Poisson summation formula (PSF) is an equation that relates the Fourier series coefficients of the periodic summation of a function to values of the function's continuous Fourier transform. The Poisson summation formula says that for sufficiently regular functions *f*:

$$\sum_n \hat{f}(n) = \sum_n f(n).$$

It has a variety of useful forms that are derived from the basic one by application of the Fourier transform's scaling and time-shifting properties. The formula has applications in engineering, physics, and number theory. The frequency-domain dual of the standard Poisson summation formula is also called the discrete-time Fourier transform.

Poisson summation is generally associated with the physics of periodic media, such as heat conduction on a circle. The fundamental solution of the heat equation on a circle is called a theta function. It is used in number theory to prove the transformation properties of theta functions, which turn out to be a type of modular form, and it is connected more generally to the theory of automorphic forms where it appears on one side of the Selberg trace formula.

Differentiation

Suppose $f(x)$ is an absolutely continuous differentiable function, and both f and its derivative f' are integrable. Then the Fourier transform of the derivative is given by:

$$f'(\xi) = 2\pi i \xi \hat{f}(\xi).$$

More generally, the Fourier transformation of the nth derivative $f^{(n)}$ is given by:

$$f^{(n)}(\xi) = (2\pi i \xi)^n \hat{f}(\xi).$$

By applying the Fourier transform and using these formulas, some ordinary differential equations can be transformed into algebraic equations, which are much easier to solve. These formulas also give rise to the rule of thumb "$f(x)$ is smooth if and only if $\hat{f}(\xi)$ quickly falls to 0 for $|\xi| \to \infty$." By using the analogous rules for the inverse Fourier transform, one can also say "$f(x)$ quickly falls to 0 for $|x| \to \infty$ if and only if $\hat{f}(\xi)$ is smooth."

Convolution Theorem

The Fourier transform translates between convolution and multiplication of functions. If $f(x)$ and $g(x)$ are integrable functions with Fourier transforms $\hat{f}(\xi)$ and $\hat{g}(\xi)$ respectively, then the Fourier transform of the convolution is given by the product of the Fourier transforms $\hat{f}(\xi)$ and $\hat{g}(\xi)$ (under other conventions for the definition of the Fourier transform a constant factor may appear).

This means that if:

$$h(x) = (f * g)(x) = \int_{-\infty}^{\infty} f(y) g(x-y) dy,$$

where $*$ denotes the convolution operation, then:

$$\hat{h}(\xi) = \hat{f}(\xi) \cdot \hat{g}(\xi).$$

In linear time invariant (LTI) system theory, it is common to interpret $g(x)$ as the impulse response of an LTI system with input $f(x)$ and output $h(x)$, since substituting the unit impulse for $f(x)$ yields $h(x) = g(x)$. In this case, $\hat{g}(\xi)$ represents the frequency response of the system.

Conversely, if $f(x)$ can be decomposed as the product of two square integrable functions $p(x)$ and $q(x)$, then the Fourier transform of $f(x)$ is given by the convolution of the respective Fourier transforms $\hat{p}(\xi)$ and $\hat{q}(\xi)$.

Cross-correlation Theorem

In an analogous manner, it can be shown that if $h(x)$ is the cross-correlation of $f(x)$ and $g(x)$:

$$h(x) = (f \star g)(x) = \int_{-\infty}^{\infty} \overline{f(y)} (x+y) dy$$

then the Fourier transform of $h(x)$ is:

$$\hat{h}(\xi) = \overline{\hat{f}(\xi)} \cdot \hat{g}(\xi).$$

As a special case, the autocorrelation of function $f(x)$ is:

$$h(x) = (f \star f)(x) = \int_{-\infty}^{\infty} \overline{f(y)} f(x+y) dy$$

for which:

$$\hat{h}(\xi) = \overline{\hat{f}(\xi)} \hat{f}(\xi) = \left| \hat{f}(\xi) \right|^2.$$

Eigenfunctions

One important choice of an orthonormal basis for $L^2(\mathbb{R})$ is given by the Hermite functions:

$$\psi_n(x) = \frac{\sqrt[4]{2}}{\sqrt{n!}} e^{-\pi x^2} \mathrm{He}_n\left(2x\sqrt{\pi}\right),$$

where $\mathrm{He}_n(x)$ are the "probabilist's" Hermite polynomials, defined as:

$$\mathrm{He}_n(x) = (-1)^n e^{\frac{x^2}{2}} \left(\frac{d}{dx}\right)^n e^{-\frac{x^2}{2}}$$

Under this convention for the Fourier transform, we have that:

$$\hat{\psi}_n(\xi) = (-i)^n \psi_n(\xi).$$

In other words, the Hermite functions form a complete orthonormal system of eigenfunctions for the Fourier transform on $L^2(\mathbb{R})$. However, this choice of eigenfunctions is not unique. There are only four different eigenvalues of the Fourier transform (±1 and ±i) and any linear combination of eigenfunctions with the same eigenvalue gives another eigenfunction. As a consequence of this, it is possible to decompose $L^2(\mathbb{R})$ as a direct sum of four spaces H_0, H_1, H_2, and H_3 where the Fourier transform acts on He_k simply by multiplication by i^k.

Since the complete set of Hermite functions provides a resolution of the identity, the Fourier transform can be represented by such a sum of terms weighted by the above eigenvalues, and these sums can be explicitly summed. This approach to define the Fourier transform was first done by Norbert Wiener. Among other properties, Hermite functions decrease exponentially fast in both frequency and time domains, and they are thus used to define a generalization of the Fourier transform, namely the fractional Fourier transform used in time-frequency analysis. In physics, this transform was introduced by Edward Condon.

Connection with the Heisenberg Group

The Heisenberg group is a certain group of unitary operators on the Hilbert space $L^2(\mathbb{R})$ of square integrable complex valued functions f on the real line, generated by the translations $(T_y f)(x) = f(x + y)$ and multiplication by $e^{2\pi i x\xi}$, $(M_\xi f)(x) = e^{2\pi i x\xi} f(x)$. These operators do not commute, as their (group) commutator is:

$$\left(M_\xi^{-1} T_y^{-1} M_\xi T_y f\right)(x) = e^{2\pi i y\xi} f(x)$$

which is multiplication by the constant (independent of x) $e^{2\pi i y\xi} \in U(1)$ (the circle group of unit modulus complex numbers). As an abstract group, the Heisenberg group is the three-dimensional Lie group of triples $(x, \xi, z) \in \mathbb{R}^2 \times U(1)$, with the group law:

$$\left(x_1, \xi_1, t_1\right) \cdot \left(x_2, \xi_2, t_2\right) = \left(x_1 + x_2, \xi_1 + \xi_2, t_1 t_2 e^{2\pi i\left(x_1\xi_1 + x_2\xi_2 + x_1\xi_2\right)}\right)$$

Denote the Heisenberg group by H_1. The above procedure describes not only the group structure, but also a standard unitary representation of H_1 on a Hilbert space, which we denote by $\rho : H_1 \to B(L^2(\mathbb{R}))$. Define the linear automorphism of $\mathbb{R}^2$ by:

$$J\begin{pmatrix} x \\ \xi \end{pmatrix} = \begin{pmatrix} -\xi \\ x \end{pmatrix}.$$

so that $J^2 = -I$. This J can be extended to a unique automorphism of H_1:

$$j(x,\xi,t) = \left(-\xi, x, te^{-2\pi i x\xi}\right).$$

According to the Stone–von Neumann theorem, the unitary representations ρ and $\rho \circ j$ are unitarily equivalent, so there is a unique intertwiner $W \in U(L^2(\mathrm{R}))$ such that:

$$\rho^\circ j = W \rho W^*.$$

This operator W is the Fourier transform.

Many of the standard properties of the Fourier transform are immediate consequences of this more general framework. For example, the square of the Fourier transform, W^2, is an intertwiner associated with $J^2 = -I$, and so we have $(W^2 f)(x) = f(-x)$ is the reflection of the original function f.

Complex Domain

The integral for the Fourier transform:

$$\hat{f}(\xi) = \int_{-\infty}^{\infty} e^{-2\pi i \xi t} f(t)dt$$

can be studied for complex values of its argument ξ. Depending on the properties of f, this might not converge off the real axis at all, or it might converge to a complex analytic function for all values of $\xi = \sigma + i\tau$, or something in between.

The Paley–Wiener theorem says that f is smooth (i.e., n-times differentiable for all positive integers n) and compactly supported if and only if $\hat{f}(\sigma + i\tau)$ is a holomorphic function for which there exists a constant $a > 0$ such that for any integer $n \geq 0$:

$$\left|\xi^n \hat{f}(\xi)\right| \leq Ce^{a|\tau|}$$

for some constant C. (In this case, f is supported on $[-a, a]$.) This can be expressed by saying that $\hat{f}$ is an entire function which is rapidly decreasing in σ (for fixed τ) and of exponential growth in τ (uniformly in σ).

(If f is not smooth, but only L^2, the statement still holds provided $n = 0$.) The space of such functions of a complex variable is called the Paley—Wiener space. This theorem has been generalised to semisimple Lie groups.

If f is supported on the half-line $t \geq 0$, then f is said to be "causal" because the impulse response function of a physically realisable filter must have this property, as no effect can precede its cause. Paley and Wiener showed that then $f^\wedge$ extends to a holomorphic function on the complex lower half-plane $\tau < 0$ which tends to zero as τ goes to infinity. The converse is false and it is not known how to characterise the Fourier transform of a causal function.

Laplace Transform

The Fourier transform $f^\wedge(\xi)$ is related to the Laplace transform $F(s)$, which is also used for the solution of differential equations and the analysis of filters.

It may happen that a function f for which the Fourier integral does not converge on the real axis at all, nevertheless has a complex Fourier transform defined in some region of the complex plane.

For example, if $f(t)$ is of exponential growth, i.e.,

$$|f(t)| < Ce^{a|t|}$$

for some constants C, $a \geq 0$, then:

$$\hat{f}(i\tau) = \int_{-\infty}^{\infty} e^{2\pi\tau t} f(t)dt,$$

convergent for all $2\pi\tau < -a$, is the two-sided Laplace transform of f.

The more usual version ("one-sided") of the Laplace transform is:

$$F(s) = \int_0^{\infty} f(t)e^{-st}\, dt.$$

If f is also causal, then:

$$\hat{f}(i\tau) = F(-2\pi\tau).$$

Thus, extending the Fourier transform to the complex domain means it includes the Laplace transform as a special case—the case of causal functions—but with the change of variable $s = 2\pi i\xi$.

Inversion

If $f^\wedge$ is complex analytic for $a \leq \tau \leq b$, then:

$$\int_{-\infty}^{\infty} \hat{f}(\sigma + ia)e^{2\pi i\xi t}\, d\sigma = \int_{-\infty}^{\infty} \hat{f}(\sigma - ib)e^{2\pi i\xi t}\, d\sigma$$

by Cauchy's integral theorem. Therefore, the Fourier inversion formula can use integration along different lines, parallel to the real axis.

Theorem: If $f(t) = 0$ for $t < 0$, and $|f(t)| < Ce^{a|t|}$ for some constants C, $a > 0$, then:

$$f(t) = \int_{-\infty}^{\infty} \hat{f}(\sigma + i\tau)e^{2\pi i\xi t}\, d\sigma,$$

for any $\tau < -a/2\pi$.

This theorem implies the Mellin inversion formula for the Laplace transformation:

$$f(t)=\frac{1}{2\pi i}\int_{b-i\infty}^{b+i\infty}F(s)e^{st}\,ds$$

for any $b > a$, where $F(s)$ is the Laplace transform of $f(t)$.

The hypotheses can be weakened, as in the results of Carleman and Hunt, to $f(t)\,e^{-at}$ being L^1, provided that f is of bounded variation in a closed neighborhood of t (cf. Dirichlet-Dini theorem), the value of f at t is taken to be the arithmetic mean of the left and right limits, and provided that the integrals are taken in the sense of Cauchy principal values.

L^2 versions of these inversion formulas are also available.

Fourier Transform on Euclidean Space

The Fourier transform can be defined in any arbitrary number of dimensions n. As with the one-dimensional case, there are many conventions. For an integrable function $f(\mathbf{x})$,

$$\hat{f}(\xi)=\mathcal{F}(f)(\xi)=\int_{\mathbb{R}^n}f(\mathbf{x})e^{-2\pi i\mathbf{x}\cdot\xi}\,d\mathbf{x}$$

where $\mathbf{x}$ and $\boldsymbol{\xi}$ are n-dimensional vectors, and $\mathbf{x}\cdot\boldsymbol{\xi}$ is the dot product of the vectors. The dot product is sometimes written as $\langle\mathbf{x}, \boldsymbol{\xi}\rangle$.

All of the basic properties listed above hold for the n-dimensional Fourier transform, as do Plancherel's and Parseval's theorem. When the function is integrable, the Fourier transform is still uniformly continuous and the Riemann–Lebesgue lemma holds.

Uncertainty Principle

Generally speaking, the more concentrated $f(x)$ is, the more spread out its Fourier transform $f^{\wedge}(\xi)$ must be. In particular, the scaling property of the Fourier transform may be seen as saying: if we squeeze a function in x, its Fourier transform stretches out in ξ. It is not possible to arbitrarily concentrate both a function and its Fourier transform.

The trade-off between the compaction of a function and its Fourier transform can be formalized in the form of an uncertainty principle by viewing a function and its Fourier transform as conjugate variables with respect to the symplectic form on the time–frequency domain: from the point of view of the linear canonical transformation, the Fourier transform is rotation by 90° in the time–frequency domain, and preserves the symplectic form.

Suppose $f(x)$ is an integrable and square-integrable function. Without loss of generality, assume that $f(x)$ is normalized:

$$\int_{-\infty}^{\infty}|f(x)|^2\,dx=1.$$

It follows from the Plancherel theorem that $f^{\wedge}(\xi)$ is also normalized.

The spread around $x = 0$ may be measured by the *dispersion about zero* defined by:

$$D_0(f) = \int_{-\infty}^{\infty} x^2 \mid f(x) \mid^2 dx.$$

In probability terms, this is the second moment of $|f(x)|^2$ about zero.

The uncertainty principle states that, if $f(x)$ is absolutely continuous and the functions $x \cdot f(x)$ and $f'(x)$ are square integrable, then:

$$D_0(f) D_0\left(\hat{f}\right) \geq \frac{1}{16\pi^2}.$$

The equality is attained only in the case:

$$f(x) = C_1 e^{-\pi \frac{x^2}{\sigma^2}}$$
$$\therefore \hat{f}(\xi) = \sigma C_1 e^{-\pi\sigma^2 \xi^2}$$

where $\sigma > 0$ is arbitrary and $C_1 = \frac{\sqrt[4]{2}}{\sqrt{\sigma}}$ so that f is L^2-normalized. In other words, where f is a (normalized) Gaussian function with variance σ^2, centered at zero, and its Fourier transform is a Gaussian function with variance σ^{-2}.

In fact, this inequality implies that:

$$\left(\int_{-\infty}^{\infty} (x - x_0)^2 \mid f(x) \mid^2 dx\right)\left(\int_{-\infty}^{\infty} (\xi - \xi_0)^2 \left|\hat{f}(\xi)\right|^2 d\xi\right) \geq \frac{1}{16\pi^2}$$

for any $x_0, \xi_0 \in \mathbb{R}$.

In quantum mechanics, the momentum and position wave functions are Fourier transform pairs, to within a factor of Planck's constant. With this constant properly taken into account, the inequality above becomes the statement of the Heisenberg uncertainty principle.

A stronger uncertainty principle is the Hirschman uncertainty principle, which is expressed as:

$$H\left(|f|^2\right) + H\left(\left|\hat{f}\right|^2\right) \geq \log\left(\frac{e}{2}\right)$$

where $H(p)$ is the differential entropy of the probability density function $p(x)$:

$$H(p) = -\int_{-\infty}^{\infty} p(x) \log(p(x)) dx$$

where the logarithms may be in any base that is consistent. The equality is attained for a Gaussian, as in the previous case.

Sine and Cosine Transforms

Fourier's original formulation of the transform did not use complex numbers, but rather sines and cosines. Statisticians and others still use this form. An absolutely integrable function *f* for which Fourier inversion holds good can be expanded in terms of genuine frequencies (avoiding negative frequencies, which are sometimes considered hard to interpret physically) λ by:

$$f(t) = \int_{\sim 0}^{\infty} \big(a(\lambda)\cos(2\pi\lambda t) + b(\lambda)\sin(2\pi\lambda t)\big) d\lambda.$$

This is called an expansion as a trigonometric integral, or a Fourier integral expansion. The coefficient functions a and b can be found by using variants of the Fourier cosine transform and the Fourier sine transform (the normalisations are, again, not standardised):

$$a(\lambda) = 2\int_{-\infty}^{\infty} f(t)\cos(2\pi\lambda t)dt$$

and

$$b(\lambda) = 2\int_{-\infty}^{\infty} f(t)\sin(2\pi\lambda t)dt.$$

Older literature refers to the two transform functions, the Fourier cosine transform, a, and the Fourier sine transform, b.

The function *f* can be recovered from the sine and cosine transform using:

$$f(t) = 2\int_{0}^{\infty}\int_{-\infty}^{\infty} f(\tau)\cos\big(2\pi\lambda(\tau - t)\big)d\tau\, d\lambda.$$

together with trigonometric identities. This is referred to as Fourier's integral formula.

Spherical Harmonics

Let the set of homogeneous harmonic polynomials of degree k on $\mathbb{R}^n$ be denoted by A_k. The set A_k consists of the solid spherical harmonics of degree k. The solid spherical harmonics play a similar role in higher dimensions to the Hermite polynomials in dimension one. Specifically, if $f(x) = e^{-\pi|x|^2}P(x)$ for some $P(x)$ in A_k, then $\hat{f}(\xi) = i^{-k} f(\xi)$. Let the set H_k be the closure in $L^2(\mathbb{R}^n)$ of linear combinations of functions of the form $f(|x|)P(x)$ where $P(x)$ is in A_k. The space $L^2(\mathbb{R}^n)$ is then a direct sum of the spaces H_k and the Fourier transform maps each space H_k to itself and is possible to characterize the action of the Fourier transform on each space H_k.

Let $f(x) = f_0(|x|)P(x)$ (with $P(x)$ in A_k), then:

$$\hat{f}(\xi) = F_0(|\xi|)P(\xi)$$

where:

$$F_0(r) = 2\pi i^{-k} r^{-\frac{n+2k-2}{2}} \int_0^{\infty} f_0(s) J_{\frac{n+2k-2}{2}}(2\pi r s) s^{\frac{n+2k}{2}} ds.$$

Here $J_{n+2k-2/2}$ denotes the Bessel function of the first kind with order $n + 2k - 2/2$. When $k = 0$ this gives a useful formula for the Fourier transform of a radial function. This is essentially the Hankel transform. Moreover, there is a simple recursion relating the cases $n + 2$ and n allowing to compute, e.g., the three-dimensional Fourier transform of a radial function from the one-dimensional one.

Restriction Problems

In higher dimensions it becomes interesting to study restriction problems for the Fourier transform. The Fourier transform of an integrable function is continuous and the restriction of this function to any set is defined. But for a square-integrable function the Fourier transform could be a general *class* of square integrable functions. As such, the restriction of the Fourier transform of an $L^2(\mathbb{R}^n)$ function cannot be defined on sets of measure 0. It is still an active area of study to understand restriction problems in L^p for $1 < p < 2$. Surprisingly, it is possible in some cases to define the restriction of a Fourier transform to a set S, provided S has non-zero curvature. The case when S is the unit sphere in $\mathbb{R}^n$ is of particular interest. In this case the Tomas–Stein restriction theorem states that the restriction of the Fourier transform to the unit sphere in $\mathbb{R}^n$ is a bounded operator on L^p provided $1 \le p \le \frac{2n+2}{n+3}$.

One notable difference between the Fourier transform in 1 dimension versus higher dimensions concerns the partial sum operator. Consider an increasing collection of measurable sets E_R indexed by $R \in (0,\infty)$: such as balls of radius R centered at the origin, or cubes of side $2R$. For a given integrable function f, consider the function f_R defined by:

$$f_R(x) = \int_{E_R} \hat{f}(\xi) e^{2\pi i x \cdot \xi}\, d\xi, \quad x \in \mathbb{R}^n.$$

Suppose in addition that $f \in L^p(\mathbb{R}^n)$. For $n = 1$ and $1 < p < \infty$, if one takes $E_R = (-R, R)$, then f_R converges to f in L^p as R tends to infinity, by the boundedness of the Hilbert transform. Naively one may hope the same holds true for $n > 1$. In the case that E_R is taken to be a cube with side length R, then convergence still holds. Another natural candidate is the Euclidean ball $E_R = \{\xi : |\xi| < R\}$. In order for this partial sum operator to converge, it is necessary that the multiplier for the unit ball be bounded in $L^p(\mathbb{R}^n)$. For $n \ge 2$ it is a celebrated theorem of Charles Fefferman that the multiplier for the unit ball is never bounded unless $p = 2$. In fact, when $p \ne 2$, this shows that not only may f_R fail to converge to f in L^p, but for some functions $f \in L^p(\mathbb{R}^n)$, f_R is not even an element of L^p.

Fourier Transform on Function Spaces

On L^p Spaces

On L^1:

The definition of the Fourier transform by the integral formula:

$$\hat{f}(\xi) = \int_{\mathbb{R}^n} f(x) e^{-2\pi i \xi \cdot x}\, dx$$

is valid for Lebesgue integrable functions f; that is, $f \in L^1(\mathbb{R}^n)$.

The Fourier transform $F: L^1(\mathbb{R}^n) \to L^\infty(\mathbb{R}^n)$ is a bounded operator. This follows from the observation that:

$$\left|\hat{f}(\xi)\right| \leq \int_{\mathbb{R}^n} |f(x)|dx,$$

which shows that its operator norm is bounded by 1. Indeed, it equals 1, which can be seen, for example, from the transform of the rect function. The image of L^1 is a subset of the space $C_0(\mathbb{R}^n)$ of continuous functions that tend to zero at infinity (the Riemann–Lebesgue lemma), although it is not the entire space. Indeed, there is no simple characterization of the image.

On L^2

Since compactly supported smooth functions are integrable and dense in $L^2(\mathbb{R}^n)$, the Plancherel theorem allows us to extend the definition of the Fourier transform to general functions in $L^2(\mathbb{R}^n)$ by continuity arguments. The Fourier transform in $L^2(\mathbb{R}^n)$ is no longer given by an ordinary Lebesgue integral, although it can be computed by an improper integral, here meaning that for an L^2 function f,

$$\hat{f}(\xi) = \lim_{R\to\infty} \int_{|x|\leq R} f(x)e^{-2\pi ix\cdot\xi}\, dx$$

where the limit is taken in the L^2 sense. (More generally, you can take a sequence of functions that are in the intersection of L^1 and L^2 and that converges to f in the L^2-norm, and define the Fourier transform of f as the L^2 -limit of the Fourier transforms of these functions).

Many of the properties of the Fourier transform in L^1 carry over to L^2, by a suitable limiting argument.

Furthermore, $F: L^2(\mathbb{R}^n) \to L^2(\mathbb{R}^n)$ is a unitary operator. For an operator to be unitary it is sufficient to show that it is bijective and preserves the inner product, so in this case these follow from the Fourier inversion theorem combined with the fact that for any $f, g \in L^2(\mathbb{R}^n)$ we have:

$$\int_{\square^n} f(x)\mathcal{F}g(x)dx = \int_{\square^n} \mathcal{F}f(x)g(x)dx.$$

In particular, the image of $L^2(\mathbb{R}^n)$ is itself under the Fourier transform.

On Other L^p

The definition of the Fourier transform can be extended to functions in $L^p(\mathbb{R}^n)$ for $1 \leq p \leq 2$ by decomposing such functions into a fat tail part in L^2 plus a fat body part in L^1. In each of these spaces, the Fourier transform of a function in $L^p(\mathbb{R}^n)$ is in $L^q(\mathbb{R}^n)$, where $q = p/p - 1$ is the Hölder conjugate of p (by the Hausdorff–Young inequality). However, except for $p = 2$, the image is not easily characterized. Further extensions become more technical. The Fourier transform of functions in L^p for the range $2 < p < \infty$ requires the study of distributions. In fact, it can be shown that there are functions in L^p with $p > 2$ so that the Fourier transform is not defined as a function.

Tempered Distributions

One might consider enlarging the domain of the Fourier transform from $L^1 + L^2$ by considering generalized functions, or distributions. A distribution on $\mathbb{R}^n$ is a continuous linear functional on the space $C_c(\mathbb{R}^n)$ of compactly supported smooth functions, equipped with a suitable topology. The strategy is then to consider the action of the Fourier transform on $C_c(\mathbb{R}^n)$ and pass to distributions by duality. The obstruction to doing this is that the Fourier transform does not map $C_c(\mathbb{R}^n)$ to $C_c(\mathbb{R}^n)$. In fact the Fourier transform of an element in $C_c(\mathbb{R}^n)$ can not vanish on an open set; The right space here is the slightly larger space of Schwartz functions. The Fourier transform is an automorphism on the Schwartz space, as a topological vector space, and thus induces an automorphism on its dual, the space of tempered distributions. The tempered distributions include all the integrable functions mentioned above, as well as well-behaved functions of polynomial growth and distributions of compact support.

For the definition of the Fourier transform of a tempered distribution, let f and g be integrable functions, and let $\hat{f}$ and $\hat{g}$ be their Fourier transforms respectively. Then the Fourier transform obeys the following multiplication formula:

$$\int_{\mathbb{R}^n} \hat{f}(x)g(x)dx = \int_{\mathbb{R}^n} f(x)\hat{g}(x)dx.$$

Every integrable function f defines (induces) a distribution T_f by the relation:

$$T_f(\varphi) = \int_{\mathbb{R}^n} f(x)\varphi(x)dx$$

for all Schwartz functions φ. So it makes sense to define Fourier transform $\hat{T}_f$ of T_f by:

$$\hat{T}_f(\varphi) = T_f(\hat{\varphi})$$

for all Schwartz functions φ. Extending this to all tempered distributions T gives the general definition of the Fourier transform.

Distributions can be differentiated and the above-mentioned compatibility of the Fourier transform with differentiation and convolution remains true for tempered distributions.

Generalizations

Fourier–Stieltjes Transform

The Fourier transform of a finite Borel measure μ on $\mathbb{R}^n$ is given by:

$$\hat{\mu}(\xi) = \int_{\mathbb{R}^n} e^{-2\pi i x\cdot\xi}\, d\mu.$$

This transform continues to enjoy many of the properties of the Fourier transform of integrable functions. One notable difference is that the Riemann–Lebesgue lemma fails for measures. In the case that $d\mu = f(x)\, dx$, then the formula above reduces to the usual definition for the Fourier transform of f. In the case that μ is the probability distribution associated to a random variable X, the Fourier–Stieltjes transform is closely related to the characteristic function, but the typical conventions in probability theory take $e^{ix\xi}$ instead of $e^{-2\pi ix\xi}$. In the case when the distribution has a probability density function this definition reduces to the Fourier transform applied to the probability density function, again with a different choice of constants.

The Fourier transform may be used to give a characterization of measures. Bochner's theorem characterizes which functions may arise as the Fourier–Stieltjes transform of a positive measure on the circle.

Furthermore, the Dirac delta function, although not a function, is a finite Borel measure. Its Fourier transform is a constant function (whose specific value depends upon the form of the Fourier transform used).

Locally Compact Abelian Groups

The Fourier transform may be generalized to any locally compact abelian group. A locally compact abelian group is an abelian group that is at the same time a locally compact Hausdorff topological space so that the group operation is continuous. If G is a locally compact abelian group, it has a translation invariant measure μ, called Haar measure. For a locally compact abelian group G, the set of irreducible, i.e. one-dimensional, unitary representations are called its characters. With its natural group structure and the topology of pointwise convergence, the set of characters $\hat{G}$ is itself a locally compact abelian group, called the Pontryagin dual of G. For a function f in $L^1(G)$, its Fourier transform is defined by:

$$\hat{f}(\xi) = \int_G \xi(x) f(x) d\mu \qquad \text{for any } \xi \in \hat{G}.$$

The Riemann–Lebesgue lemma holds in this case; $\hat{f}(\xi)$ is a function vanishing at infinity on $\hat{G}$.

Gelfand Transform

The Fourier transform is also a special case of Gelfand transform. In this particular context, it is closely related to the Pontryagin duality map defined above.

Given an abelian locally compact Hausdorff topological group G, as before we consider space $L^1(G)$, defined using a Haar measure. With convolution as multiplication, $L^1(G)$ is an abelian Banach algebra. It also has an involution given by:

$$f^*(g) = \overline{f\left(g^{-1}\right)}.$$

Taking the completion with respect to the largest possibly C^*-norm gives its enveloping C^*-algebra, called the group C^*-algebra $C^*(G)$ of G. (Any C^*-norm on $L^1(G)$ is bounded by the L^1 norm, therefore their supremum exists.)

Given any abelian C^*-algebra A, the Gelfand transform gives an isomorphism between A and $C_0(A^\wedge)$, where $A^\wedge$ is the multiplicative linear functionals, i.e. one-dimensional representations, on A with the weak-* topology. The map is simply given by:

$$a \mapsto (\varphi \mapsto \varphi(a))$$

It turns out that the multiplicative linear functionals of $C^*(G)$, after suitable identification, are exactly the characters of G, and the Gelfand transform, when restricted to the dense subset $L^1(G)$ is the Fourier–Pontryagin transform.

Compact Non-abelian Groups

The Fourier transform can also be defined for functions on a non-abelian group, provided that the group is compact. Removing the assumption that the underlying group is abelian, irreducible unitary representations need not always be one-dimensional. This means the Fourier transform on a non-abelian group takes values as Hilbert space operators. The Fourier transform on compact groups is a major tool in representation theory and non-commutative harmonic analysis.

Let G be a compact Hausdorff topological group. Let Σ denote the collection of all isomorphism classes of finite-dimensional irreducible unitary representations, along with a definite choice of representation $U^{(\sigma)}$ on the Hilbert space H_σ of finite dimension d_σ for each $\sigma \in \Sigma$. If μ is a finite Borel measure on G, then the Fourier–Stieltjes transform of μ is the operator on H_σ defined by:

$$\langle \hat{\mu}\xi, \eta \rangle_{H_\sigma} = \int_G \langle \bar{U}_g^{(\sigma)}\xi, \eta \rangle d\mu(g)$$

where $\bar{U}^{(\sigma)}$ is the complex-conjugate representation of $U^{(\sigma)}$ acting on H_σ. If μ is absolutely continuous with respect to the left-invariant probability measure λ on G, represented as:

$$d\mu = f\, d\lambda$$

for some $f \in L^1(\lambda)$, one identifies the Fourier transform of f with the Fourier–Stieltjes transform of μ.

The mapping:

$$\mu \mapsto \hat{\mu}$$

defines an isomorphism between the Banach space $M(G)$ of finite Borel measures and a closed subspace of the Banach space $\mathbf{C}_\infty(\Sigma)$ consisting of all sequences $E = (E_\sigma)$ indexed by Σ of (bounded) linear operators $E_\sigma : H_\sigma \to H_\sigma$ for which the norm:

$$\| E \| = \sup_{\sigma \in \Sigma} \| E_\sigma \|$$

is finite. The "convolution theorem" asserts that, furthermore, this isomorphism of Banach spaces is in fact an isometric isomorphism of C* algebras into a subspace of $\mathbf{C}_\infty(\Sigma)$. Multiplication on $M(G)$ is given by convolution of measures and the involution * defined by:

$$f^*(g) = \overline{f\left(g^{-1}\right)},$$

and $\mathbf{C}_\infty(\Sigma)$ has a natural C^*-algebra structure as Hilbert space operators.

The Peter–Weyl theorem holds, and a version of the Fourier inversion formula (Plancherel's theorem) follows: if $f \in L^2(G)$, then:

$$f(g) = \sum_{\sigma \in \Sigma} d_\sigma \operatorname{tr}\left(\hat{f}(\sigma) U_g^{(\sigma)}\right)$$

where the summation is understood as convergent in the L^2 sense.

The generalization of the Fourier transform to the noncommutative situation has also in part contributed to the development of noncommutative geometry. In this context, a categorical generalization of the Fourier transform to noncommutative groups is Tannaka–Krein duality, which replaces the group of characters with the category of representations. However, this loses the connection with harmonic functions.

Alternatives

In signal processing terms, a function (of time) is a representation of a signal with perfect *time resolution*, but no frequency information, while the Fourier transform has perfect *frequency resolution*, but no time information: the magnitude of the Fourier transform at a point is how much frequency content there is, but location is only given by phase (argument of the Fourier transform at a point), and standing waves are not localized in time – a sine wave continues out to infinity, without decaying. This limits the usefulness of the Fourier transform for analyzing signals that are localized in time, notably transients, or any signal of finite extent.

As alternatives to the Fourier transform, in time-frequency analysis, one uses time-frequency transforms or time-frequency distributions to represent signals in a form that has some time information and some frequency information – by the uncertainty principle, there is a trade-off between these. These can be generalizations of the Fourier transform, such as the short-time Fourier transform or fractional Fourier transform, or other functions to represent signals, as in wavelet transforms and chirplet transforms, with the wavelet analog of the (continuous) Fourier transform being the continuous wavelet transform.

Applications

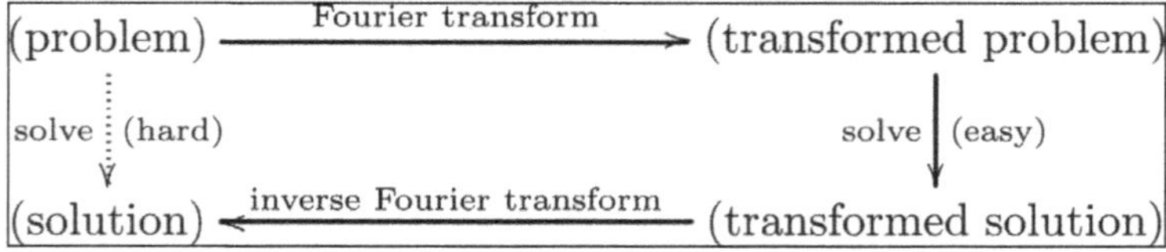

Some problems, such as certain differential equations, become easier to solve when the Fourier transform is applied. In that case the solution to the original problem is recovered using the inverse Fourier transform.

Perhaps the most important use of the Fourier transformation is to solve partial differential equations. Many of the equations of the mathematical physics of the nineteenth century can be treated this way. Fourier studied the heat equation, which in one dimension and in dimensionless units is:

$$\frac{\partial^2 y(x,t)}{\partial^2 x} = \frac{\partial y(x,t)}{\partial t}.$$

The example we will give, a slightly more difficult one, is the wave equation in one dimension,

$$\frac{\partial^2 y(x,t)}{\partial^2 x} = \frac{\partial^2 y(x,t)}{\partial^2 t}.$$

As usual, the problem is not to find a solution: there are infinitely many. The problem is that of the so-called "boundary problem": find a solution which satisfies the "boundary conditions":

$$y(x,0) = f(x), \qquad \frac{\partial y(x,0)}{\partial t} = g(x).$$

Here, f and g are given functions. For the heat equation, only one boundary condition can be required (usually the first one). But for the wave equation, there are still infinitely many solutions y which satisfy the first boundary condition. But when one imposes both conditions, there is only one possible solution.

It is easier to find the Fourier transform $\hat{y}$ of the solution than to find the solution directly. This is because the Fourier transformation takes differentiation into multiplication by the variable, and so a partial differential equation applied to the original function is transformed into multiplication by polynomial functions of the dual variables applied to the transformed function. After $\hat{y}$ is determined, we can apply the inverse Fourier transformation to find y.

Fourier's method is as follows. First, note that any function of the forms:

$$\cos\big(2\pi\xi(x \pm t)\big) \text{ or } \sin\big(2\pi\xi(x \pm t)\big)$$

satisfies the wave equation. These are called the elementary solutions.

Second, note that therefore any integral:

$$y(x,t) = \int_0^\infty a_+(\xi)\cos\big(2\pi\xi(x+t)\big) + a_-(\xi)\cos\big(2\pi\xi(x-t)\big) + b_+(\xi)\sin\big(2\pi\xi(x+t)\big) + b_-(\xi)\sin\big(2\pi\xi(x-t)\big)d\xi$$

(for arbitrary a_+, a_-, b_+, b_-) satisfies the wave equation. (This integral is just a kind of continuous linear combination, and the equation is linear.)

Now this resembles the formula for the Fourier synthesis of a function. In fact, this is the real inverse Fourier transform of $a_\pm$ and $b_\pm$ in the variable x.

The third step is to examine how to find the specific unknown coefficient functions $a_\pm$ and $b_\pm$ that will lead to y satisfying the boundary conditions. We are interested in the values of these solutions at $t = 0$. So we will set $t = 0$. Assuming that the conditions needed for Fourier inversion are satisfied, we can then find the Fourier sine and cosine transforms (in the variable x) of both sides and obtain:

$$2\int_{-\infty}^{\infty} y(x,0)\cos(2\pi\xi x)dx = a_+ + a_-$$

and

$$2\int_{-\infty}^{\infty} y(x,0)\sin(2\pi\xi x)dx = b_+ + b_-.$$

Similarly, taking the derivative of y with respect to t and then applying the Fourier sine and cosine transformations yields,

$$2\int_{-\infty}^{\infty}\frac{\partial y(u,0)}{\partial t}\sin(2\pi\xi x)dx=(2\pi\xi)\left(-a_{+}+a_{-}\right)$$

and

$$2\int_{-\infty}^{\infty}\frac{\partial y(u,0)}{\partial t}\cos(2\pi\xi x)dx=(2\pi\xi)\left(b_{+}-b_{-}\right).$$

These are four linear equations for the four unknowns $a_{\pm}$ and $b_{\pm}$, in terms of the Fourier sine and cosine transforms of the boundary conditions, which are easily solved by elementary algebra, provided that these transforms can be found.

In summary, we chose a set of elementary solutions, parametrised by ξ, of which the general solution would be a (continuous) linear combination in the form of an integral over the parameter ξ. But this integral was in the form of a Fourier integral. The next step was to express the boundary conditions in terms of these integrals, and set them equal to the given functions f and g. But these expressions also took the form of a Fourier integral because of the properties of the Fourier transform of a derivative. The last step was to exploit Fourier inversion by applying the Fourier transformation to both sides, thus obtaining expressions for the coefficient functions $a_{\pm}$ and $b_{\pm}$ in terms of the given boundary conditions f and g.

From a higher point of view, Fourier's procedure can be reformulated more conceptually. Since there are two variables, we will use the Fourier transformation in both x and t rather than operate as Fourier did, who only transformed in the spatial variables. Note that $\hat{y}$ must be considered in the sense of a distribution since $y(x, t)$ is not going to be L^1: as a wave, it will persist through time and thus is not a transient phenomenon. But it will be bounded and so its Fourier transform can be defined as a distribution. The operational properties of the Fourier transformation that are relevant to this equation are that it takes differentiation in x to multiplication by $2\pi i\xi$ and differentiation with respect to t to multiplication by $2\pi if$ where f is the frequency. Then the wave equation becomes an algebraic equation in $\hat{y}$:

$$\xi^2\hat{y}(\xi,f)=f^2\hat{y}(\xi,f).$$

This is equivalent to requiring $\hat{y}(\xi,f) = 0$ unless $\xi = \pm f$. Right away, this explains why the choice of elementary solutions we made earlier worked so well: obviously $\hat{f} = \delta(\xi \pm f)$ will be solutions. Applying Fourier inversion to these delta functions, we obtain the elementary solutions we picked earlier. But from the higher point of view, one does not pick elementary solutions, but rather considers the space of all distributions which are supported on the (degenerate) conic $\xi^2 - f^2 = 0$.

We may as well consider the distributions supported on the conic that are given by distributions of one variable on the line $\xi = f$ plus distributions on the line $\xi = -f$ as follows: if ϕ is any test function:

$$\iint \hat{y}\phi(\xi,f)d\xi df=\int s_{+}\phi(\xi,\xi)d\xi+\int s_{-}\phi(\xi,-\xi)d\xi,$$

where s_{+}, and s_{-}, are distributions of one variable.

Then Fourier inversion gives, for the boundary conditions, something very similar to what we had more concretely above (put $\phi(\xi,f) = e^{2\pi i(x\xi+tf)}$, which is clearly of polynomial growth):

$$y(x,0) = \int\{s_+(\xi) + s_-(\xi)\}e^{2\pi i\xi x+0}\, d\xi$$

and

$$\frac{\partial y(x,0)}{\partial t} = \int\{s_+(\xi) - s_-(\xi)\}2\pi i\xi e^{2\pi i\xi x+0}\, d\xi.$$

Now, as before, applying the one-variable Fourier transformation in the variable x to these functions of x yields two equations in the two unknown distributions $s_\pm$ (which can be taken to be ordinary functions if the boundary conditions are L^1 or L^2).

From a calculational point of view, the drawback of course is that one must first calculate the Fourier transforms of the boundary conditions, then assemble the solution from these, and then calculate an inverse Fourier transform. Closed form formulas are rare, except when there is some geometric symmetry that can be exploited, and the numerical calculations are difficult because of the oscillatory nature of the integrals, which makes convergence slow and hard to estimate. For practical calculations, other methods are often used.

The twentieth century has seen the extension of these methods to all linear partial differential equations with polynomial coefficients, and by extending the notion of Fourier transformation to include Fourier integral operators, some non-linear equations as well.

Fourier Transform Spectroscopy

The Fourier transform is also used in nuclear magnetic resonance (NMR) and in other kinds of spectroscopy, e.g. infrared (FTIR). In NMR an exponentially shaped free induction decay (FID) signal is acquired in the time domain and Fourier-transformed to a Lorentzian line-shape in the frequency domain. The Fourier transform is also used in magnetic resonance imaging (MRI) and mass spectrometry.

Quantum Mechanics

The Fourier transform is useful in quantum mechanics in two different ways. To begin with, the basic conceptual structure of Quantum Mechanics postulates the existence of pairs of complementary variables, connected by the Heisenberg uncertainty principle. For example, in one dimension, the spatial variable q of, say, a particle, can only be measured by the quantum mechanical "position operator" at the cost of losing information about the momentum p of the particle. Therefore, the physical state of the particle can either be described by a function, called "the wave function", of q or by a function of p but not by a function of both variables. The variable p is called the conjugate variable to q. In Classical Mechanics, the physical state of a particle (existing in one dimension, for simplicity of exposition) would be given by assigning definite values to both p and q simultaneously. Thus, the set of all possible physical states is the two-dimensional real vector space with a p-axis and a q-axis called the phase space.

In contrast, quantum mechanics chooses a polarisation of this space in the sense that it picks a subspace of one-half the dimension, for example, the q-axis alone, but instead of considering only points, takes the set of all complex-valued "wave functions" on this axis. Nevertheless, choosing the p-axis is an equally valid polarisation, yielding a different representation of the set of possible physical states of the particle which is related to the first representation by the Fourier transformation:

$$\phi(p) = \int \psi(q) e^{2\pi i \frac{pq}{h}} dq.$$

Physically realisable states are L^2, and so by the Plancherel theorem, their Fourier transforms are also L^2. Note that since q is in units of distance and p is in units of momentum, the presence of Planck's constant in the exponent makes the exponent dimensionless, as it should be.

Therefore, the Fourier transform can be used to pass from one way of representing the state of the particle, by a wave function of position, to another way of representing the state of the particle: by a wave function of momentum. Infinitely many different polarisations are possible, and all are equally valid. Being able to transform states from one representation to another is sometimes convenient.

The other use of the Fourier transform in both quantum mechanics and quantum field theory is to solve the applicable wave equation. In non-relativistic quantum mechanics, Schrödinger's equation for a time-varying wave function in one-dimension, not subject to external forces, is:

$$\frac{\partial^2}{\partial x^2} \psi(x,t) = i \frac{h}{2\pi} \frac{\partial}{\partial t} \psi(x,t).$$

This is the same as the heat equation except for the presence of the imaginary unit i. Fourier methods can be used to solve this equation.

In the presence of a potential, given by the potential energy function $V(x)$, the equation becomes:

$$\frac{\partial^2}{\partial x^2} \psi(x,t) + V(x)\psi(x,t) = i \frac{h}{2\pi} \frac{\partial}{\partial t} \psi(x,t).$$

The "elementary solutions", as we referred to them above, are the so-called "stationary states" of the particle, and Fourier's algorithm, as described above, can still be used to solve the boundary value problem of the future evolution of ψ given its values for $t = 0$. Neither of these approaches is of much practical use in quantum mechanics. Boundary value problems and the time-evolution of the wave function is not of much practical interest: it is the stationary states that are most important.

In relativistic quantum mechanics, Schrödinger's equation becomes a wave equation as was usual in classical physics, except that complex-valued waves are considered. A simple example, in the absence of interactions with other particles or fields, is the free one-dimensional Klein–Gordon–Schrödinger–Fock equation, this time in dimensionless units:

$$\left(\frac{\partial^2}{\partial x^2} + 1 \right) \psi(x,t) = \frac{\partial^2}{\partial t^2} \psi(x,t).$$

This is, from the mathematical point of view, the same as the wave equation of classical physics solved above (but with a complex-valued wave, which makes no difference in the methods). This is of great use in quantum field theory: each separate Fourier component of a wave can be treated as a separate harmonic oscillator and then quantized, a procedure known as "second quantization". Fourier methods have been adapted to also deal with non-trivial interactions.

Signal Processing

The Fourier transform is used for the spectral analysis of time-series. The subject of statistical signal processing does not, however, usually apply the Fourier transformation to the signal itself. Even if a real signal is indeed transient, it has been found in practice advisable to model a signal by a function (or, alternatively, a stochastic process) which is stationary in the sense that its characteristic properties are constant over all time. The Fourier transform of such a function does not exist in the usual sense, and it has been found more useful for the analysis of signals to instead take the Fourier transform of its autocorrelation function.

The autocorrelation function R of a function f is defined by:

$$R_f(\tau) = \lim_{T\to\infty} \frac{1}{2T} \int_{-T}^{T} f(t) f(t+\tau) dt.$$

This function is a function of the time-lag τ elapsing between the values of f to be correlated.

For most functions f that occur in practice, R is a bounded even function of the time-lag τ and for typical noisy signals it turns out to be uniformly continuous with a maximum at $\tau = 0$.

The autocorrelation function, more properly called the autocovariance function unless it is normalized in some appropriate fashion, measures the strength of the correlation between the values of f separated by a time lag. This is a way of searching for the correlation of f with its own past. It is useful even for other statistical tasks besides the analysis of signals. For example, if $f(t)$ represents the temperature at time t, one expects a strong correlation with the temperature at a time lag of 24 hours.

It possesses a Fourier transform:

$$P_f(\xi) = \int_{-\infty}^{\infty} R_f(\tau) e^{-2\pi i \xi \tau} d\tau.$$

This Fourier transform is called the power spectral density function of f. (Unless all periodic components are first filtered out from f, this integral will diverge, but it is easy to filter out such periodicities.)

The power spectrum, as indicated by this density function P, measures the amount of variance contributed to the data by the frequency ξ. In electrical signals, the variance is proportional to the average power (energy per unit time), and so the power spectrum describes how much the different frequencies contribute to the average power of the signal. This process is called the spectral analysis of time-series and is analogous to the usual analysis of variance of data that is not a time-series (ANOVA).

Knowledge of which frequencies are "important" in this sense is crucial for the proper design of filters and for the proper evaluation of measuring apparatuses. It can also be useful for the scientific analysis of the phenomena responsible for producing the data.

The power spectrum of a signal can also be approximately measured directly by measuring the average power that remains in a signal after all the frequencies outside a narrow band have been filtered out.

Spectral analysis is carried out for visual signals as well. The power spectrum ignores all phase relations, which is good enough for many purposes, but for video signals other types of spectral analysis must also be employed, still using the Fourier transform as a tool.

Other Notations

Other common notations for $\hat{f}(\xi)$ include:

$$\tilde{f}(\xi), \tilde{f}(\omega), F(\xi), \mathcal{F}(f)(\xi), (\mathcal{F}f)(\xi), \mathcal{F}(f), \mathcal{F}(\omega), F(\omega), \mathcal{F}(j\omega), \mathcal{F}\{f\}, \mathcal{F}(f(t)), \mathcal{F}\{f(t)\}.$$

Denoting the Fourier transform by a capital letter corresponding to the letter of function being transformed (such as $f(x)$ and $F(\xi)$) is especially common in the sciences and engineering. In electronics, omega (ω) is often used instead of ξ due to its interpretation as angular frequency, sometimes it is written as $F(j\omega)$, where j is the imaginary unit, to indicate its relationship with the Laplace transform, and sometimes it is written informally as $F(2\pi f)$ in order to use ordinary frequency. In some contexts such as particle physics, the same symbol f may be used for both for a function as well as it Fourier transform, with the two only distinguished by their argument: $f(k_1 + k_2)$ would refer to the Fourier transform because of the momentum argument, while $f(x_0 + \pi\vec{r})$ would refer to the original function because of the positional argument. Although tildes may be used as in $\tilde{f}$ to indicate Fourier transforms, tildes may also be used to indicate a modification of a quantity with a more Lorentz invariant form, such as $\tilde{d}k = \frac{dk}{(2\pi)^3 2\omega}$, so care must be taken.

The interpretation of the complex function $\hat{f}(\xi)$ may be aided by expressing it in polar coordinate form:

$$\hat{f}(\xi) = A(\xi)e^{i\varphi(\xi)}$$

in terms of the two real functions $A(\xi)$ and $\varphi(\xi)$ where:

$$A(\xi) = \left|\hat{f}(\xi)\right|,$$

is the amplitude and,

$$\varphi(\xi) = \arg\left(\hat{f}(\xi)\right),$$

is the phase.

Then the inverse transform can be written:

$$f(x) = \int_{-\infty}^{\infty} A(\xi)\, e^{i(2\pi\xi x + \varphi(\xi))}\, d\xi,$$

which is a recombination of all the frequency components of $f(x)$. Each component is a complex sinusoid of the form $e^{2\pi i x \xi}$ whose amplitude is $A(\xi)$ and whose initial phase angle (at $x = 0$) is $\varphi(\xi)$.

The Fourier transform may be thought of as a mapping on function spaces. This mapping is here denoted F and $F(f)$ is used to denote the Fourier transform of the function f. This mapping is linear, which means that F can also be seen as a linear transformation on the function space and implies that the standard notation in linear algebra of applying a linear transformation to a vector (here the function f) can be used to write Ff instead of $F(f)$. Since the result of applying the Fourier transform is again a function, we can be interested in the value of this function evaluated at the value ξ for its variable, and this is denoted either as $Ff(\xi)$ or as $(Ff)(\xi)$. Notice that in the former case, it is implicitly understood that F is applied first to f and then the resulting function is evaluated at ξ, not the other way around.

In mathematics and various applied sciences, it is often necessary to distinguish between a function f and the value of f when its variable equals x, denoted $f(x)$. This means that a notation like $F(f(x))$ formally can be interpreted as the Fourier transform of the values of f at x. Despite this flaw, the previous notation appears frequently, often when a particular function or a function of a particular variable is to be transformed. For example,

$$\mathcal{F}\big(\operatorname{rect}(x)\big) = \operatorname{sinc}(\xi)$$

is sometimes used to express that the Fourier transform of a rectangular function is a sinc function:

$$\mathcal{F}(f(x+x_0)) = \mathcal{F}(f(x))e^{2\pi i \xi x_0}$$

is used to express the shift property of the Fourier transform.

Notice, that the last example is only correct under the assumption that the transformed function is a function of x, not of x_0.

Other Conventions

The Fourier transform can also be written in terms of angular frequency:

$$\omega = 2\pi\xi,$$

whose units are radians per second.

The substitution $\xi = \dfrac{\omega}{2\pi}$ into the formulas above produces this convention:

$$\hat{f}(\omega) = \int_{\mathbb{R}^n} f(x)e^{-i\omega\cdot x}\,dx.$$

Under this convention, the inverse transform becomes:

$$f(x) = \frac{1}{(2\pi)^n}\int_{\mathbb{R}^n} \hat{f}(\omega)e^{i\omega\cdot x}\,d\omega.$$

Unlike the convention followed here, when the Fourier transform is defined this way, it is no longer a unitary transformation on $L^2(\mathbb{R}^n)$. There is also less symmetry between the formulas for the Fourier transform and its inverse.

Another convention is to split the factor of $(2\pi)^n$ evenly between the Fourier transform and its inverse, which leads to definitions:

$$\hat{f}(\omega) = \frac{1}{(2\pi)^{\frac{n}{2}}} \int_{\mathbb{R}^n} f(x) e^{-i\omega \cdot x}\, dx,$$

$$f(x) = \frac{1}{(2\pi)^{\frac{n}{2}}} \int_{\mathbb{R}^n} \hat{f}(\omega) e^{i\omega \cdot x}\, d\omega.$$

Under this convention, the Fourier transform is again a unitary transformation on $L^2(\mathbb{R}^n)$. It also restores the symmetry between the Fourier transform and its inverse.

Variations of all three conventions can be created by conjugating the complex-exponential kernel of both the forward and the reverse transform. The signs must be opposites. Other than that, the choice is (again) a matter of convention.

Summary of popular forms of the Fourier transform, one-dimensional		
Ordinary frequency ξ (Hz)	Unitary	$\hat{f}_1(\xi) \overset{\text{def}}{=} \int_{-\infty}^{\infty} f(x) \cdot e^{-2\pi i x \cdot \xi}\, dx = \sqrt{2\pi} \cdot \hat{f}_2(2\pi\xi) = \hat{f}_3(2\pi\xi)$ $f(x) = \int_{-\infty}^{\infty} \hat{f}_1(\xi) \cdot e^{2\pi i x \cdot \xi}\, d\xi$
Angular frequency ω (rad/s)	Unitary	$\hat{f}_2(\omega) \overset{\text{def}}{=} \frac{1}{\sqrt{2\pi}} \int_{-\infty}^{\infty} f(x) \cdot e^{-i\omega \cdot x}\, dx = \frac{1}{\sqrt{2\pi}} \cdot \hat{f}_1\left(\frac{\omega}{2\pi}\right) = \frac{1}{\sqrt{2\pi}} \cdot \hat{f}_3(\omega)$ $f(x) = \frac{1}{\sqrt{2\pi}} \int_{-\infty}^{\infty} \hat{f}_2(\omega) \cdot e^{i\omega \cdot x}\, d\omega$
	Non-unitary	$\hat{f}_3(\omega) \overset{\text{def}}{=} \int_{-\infty}^{\infty} f(x) \cdot e^{-i\omega \cdot x}\, dx = \hat{f}_1\left(\frac{\omega}{2\pi}\right) = \sqrt{2\pi} \cdot \hat{f}_2(\omega)$ $f(x) = \frac{1}{2\pi} \int_{-\infty}^{\infty} \hat{f}_3(\omega) \cdot e^{i\omega \cdot x}\, d\omega$

Generalization for n-dimensional functions		
Ordinary frequency ξ (Hz)	Unitary	$\hat{f}_1(\xi) \overset{\text{def}}{=} \int_{\mathbb{R}^n} f(x) e^{-2\pi i x \cdot \xi}\, dx = (2\pi)^{\frac{n}{2}} \hat{f}_2(2\pi\xi) = \hat{f}_3(2\pi\xi)$ $f(x) = \int_{\mathbb{R}^n} \hat{f}_1(\xi) e^{2\pi i x \cdot \xi}\, d\xi$

Angular frequency ω (rad/s)	Unitary	$\hat{f}_2(\omega) \overset{\text{def}}{=} \frac{1}{(2\pi)^{\frac{n}{2}}}\int_{\mathbb{R}^n} f(x)e^{-i\omega\cdot x}\,dx = \frac{1}{(2\pi)^{\frac{n}{2}}}\hat{f}_1\left(\frac{\omega}{2\pi}\right) = \frac{1}{(2\pi)^{\frac{n}{2}}}\hat{f}_3(\omega)$ $f(x) = \frac{1}{(2\pi)^{\frac{n}{2}}}\int_{\mathbb{R}^n}\hat{f}_2(\omega)e^{i\omega\cdot x}\,d\omega$
	Non-unitary	$\hat{f}_3(\omega) \overset{\text{def}}{=} \int_{\mathbb{R}^n} f(x)e^{-i\omega\cdot x}\,dx = \hat{f}_1\left(\frac{\omega}{2\pi}\right) = (2\pi)^{\frac{n}{2}}\hat{f}_2(\omega)$ $f(x) = \frac{1}{(2\pi)^n}\int_{\mathbb{R}^n}\hat{f}_3(\omega)e^{i\omega\cdot x}\,d\omega$

The characteristic function of a random variable is the same as the Fourier–Stieltjes transform of its distribution measure, but in this context it is typical to take a different convention for the constants. Typically characteristic function is defined:

$$\left(e^{it\cdot X}\right) = \int e^{it\cdot x}\,d\mu_X(x).$$

As in the case of the "non-unitary angular frequency" convention above, the factor of 2π appears in neither the normalizing constant nor the exponent. Unlike any of the conventions appearing above, this convention takes the opposite sign in the exponent.

Computation Methods

The appropriate computation method largely depends how the original mathematical function is represented and the desired form of the output function.

Since the fundamental definition of a Fourier transform is an integral, functions that can be expressed as closed-form expressions are commonly computed by working the integral analytically to yield a closed-form expression in the Fourier transform conjugate variable as the result.

Many computer algebra systems such as Matlab and Mathematica that are capable of symbolic integration are capable of computing Fourier transforms analytically. For example, to compute the Fourier transform of $f(t) = \cos(6\pi t)\, e^{-\pi t^2}$ one might enter the command integrate cos(6*pi*t) exp(−pi*t^2) exp(-i*2*pi*f*t) from -inf to inf into Wolfram Alpha.

Numerical Integration of Closed-form Functions

If the input function is in closed-form and the desired output function is a series of ordered pairs (for example a table of values from which a graph can be generated) over a specified domain, then the Fourier transform can be generated by numerical integration at each value of the Fourier conjugate variable (frequency, for example) for which a value of the output variable is desired. Note that this method requires computing a separate numerical integration for each value of frequency for which a value of the Fourier transform is desired. The numerical integration approach works on a much broader class of functions than the analytic approach, because it yields results for functions that do not have closed form Fourier transform integrals.

Numerical Integration of a Series of Ordered Pairs

If the input function is a series of ordered pairs (for example, a time series from measuring an output variable repeatedly over a time interval) then the output function must also be a series of ordered pairs (for example, a complex number vs. frequency over a specified domain of frequencies), unless certain assumptions and approximations are made allowing the output function to be approximated by a closed-form expression. In the general case where the available input series of ordered pairs are assumed be samples representing a continuous function over an interval (amplitude vs. time, for example), the series of ordered pairs representing the desired output function can be obtained by numerical integration of the input data over the available interval at each value of the Fourier conjugate variable (frequency, for example) for which the value of the Fourier transform is desired.

Explicit numerical integration over the ordered pairs can yield the Fourier transform output value for any desired value of the conjugate Fourier transform variable (frequency, for example), so that a spectrum can be produced at any desired step size and over any desired variable range for accurate determination of amplitudes, frequencies, and phases corresponding to isolated peaks. Unlike limitations in DFT and FFT methods, explicit numerical integration can have any desired step size and compute the Fourier transform over any desired range of the conjugate Fourier transform variable (for example, frequency).

Discrete Fourier Transforms and Fast Fourier Transforms

If the ordered pairs representing the original input function are equally spaced in their input variable (for example, equal time steps), then the Fourier transform is known as a discrete Fourier transform (DFT), which can be computed either by explicit numerical integration, by explicit evaluation of the DFT definition, or by fast Fourier transform (FFT) methods. In contrast to explicit integration of input data, use of the DFT and FFT methods produces Fourier transforms described by ordered pairs of step size equal to the reciprocal of the original sampling interval. For example, if the input data is sampled every 10 seconds, the output of DFT and FFT methods will have a 0.1 Hz frequency spacing.

Fourier Transform of Standards Signals

Impulse Function δ (t)

Given $x(t) = \delta(t)$,

$$\delta(t) = \begin{cases} 1 \text{ for } t = 0 \\ 0 \text{ for } t \neq 0 \end{cases}$$

Then

$$X(\grave{u}) = \int_{-\infty}^{\infty} x(t)e^{-j\grave{u}t}\,dt = \int_{-\infty}^{\infty} \delta(t)e^{-j\grave{u}t}\,dt \quad = \quad e^{-j\grave{u}t}\Big|_{t=0} \quad = \quad 1$$

$$F[\delta(t)] = 1 \qquad \text{or } \delta(t) \overset{FT}{\leftrightarrow} 1$$

Hence, the Fourier Transform of a unit impulse function is unity.

$$|X(\omega)| = 1 \quad \text{for all } \omega$$

$$\lfloor X(\omega) \rfloor = 0 \quad \text{for all } \omega$$

The impulse function with its magnitude and phase spectra are shown in below figure:

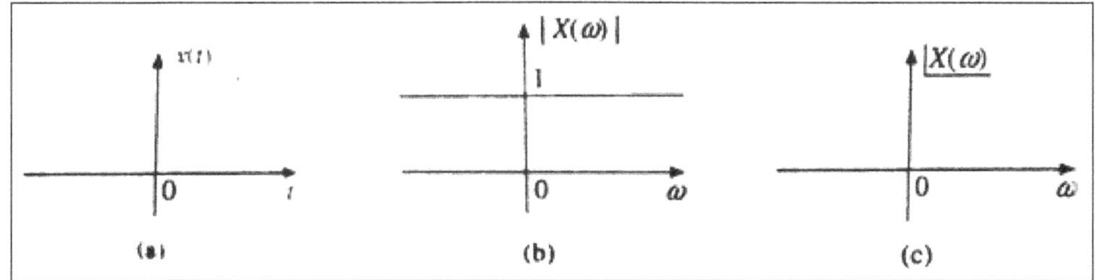

Similarly,

$$F[\delta(t-t_o)] = \int_{-\infty}^{\infty} \delta(t-t_o) e^{-j\omega t} dt = e^{-j\omega t_0} \text{ i. e. ä}(t-t_o) \overset{FT}{\leftrightarrow} e^{-j\omega t_0}$$

Single Sided Real Exponential Function $e^{-at}u(t)$

Given $x(t) = e^{-at}u(t)$, $\quad u(t) = \begin{cases} 1 & \text{for } t \geq 0 \\ 0 & \text{for } t < 0 \end{cases}$

Then,

$$x(\omega) = \int_{-\infty}^{\infty} x(t) e^{-j\omega t} dt = \int_{-\infty}^{\infty} e^{-at} u(t) e^{-j\omega t} dt$$

$$= \int_{0}^{\infty} e^{-at} e^{-j\omega t} dt = \int_{0}^{\infty} e^{-(a+j\omega)t} dt = \left[\frac{e^{(-a+j\omega)t}}{-(a+j\omega)} \right]_0^{\infty} = \frac{e^{-at} - e^0}{-(a+j\omega)}$$

$$= \frac{0-1}{-(a+j\omega)} = \frac{1}{a+j\omega}$$

$$\therefore F[e^{-at}u(t)] = \frac{1}{a+j\omega} \quad \text{or } e^{-at}u(t) \overset{FT}{\leftrightarrow} \frac{1}{a+j\omega}$$

Now,

$$X(\omega) = \frac{1}{a+j\omega} = \frac{a-j\omega}{(a+\omega)(a-\omega)}$$

$$= \frac{a - j\omega}{a^2 + \omega^2} = \frac{a}{a^2 + \omega^2} - j\frac{\omega}{a^2 + \omega^2} = \frac{1}{\sqrt{a^2 + \omega^2}}\left\lfloor -\tan^{-1}\frac{\omega}{a}\right\rfloor$$

$$\therefore |X(\omega)| = \frac{1}{\sqrt{a^2 + \omega^2}}, \left\lfloor X(\omega) = -\tan^{-1}\frac{\omega}{a} \text{for all}\ \omega\right\rfloor$$

Figure shows the single-sided exponential function with its magnitude and phase spectra.

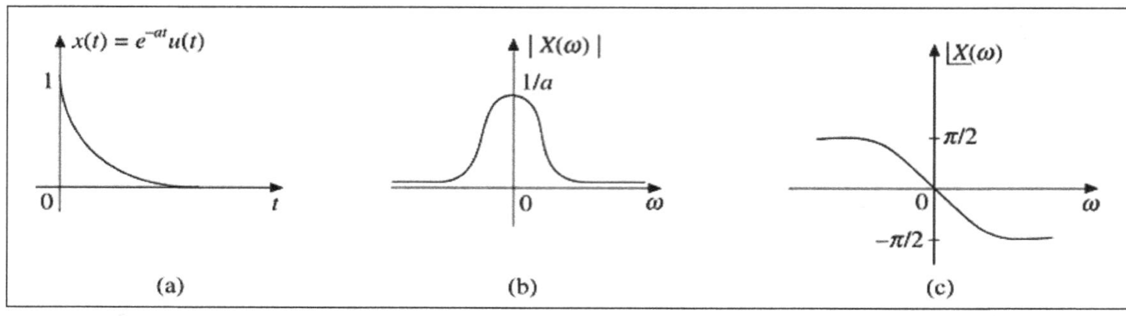

Double Sided Real Exponential Function $e^{-a|t|}$

Given $x(t) = e^{-a|t|}$

$$\therefore x(t) = e^{-a|t|} = \begin{cases} e^{-a(-t)} = e^{at} \text{for} t \le 0 \\ e^{-at} = e^{-at} \text{for} t \ge 0 \end{cases}$$

$$= e - a(-t) + e^{-at}u(t)$$

$$= e^{at}u(-t) + e^{-at}u(t)$$

$$x(\omega) = \int_{-\infty}^{\infty} x(t)\, e^{-j\omega t} dt$$

$$= \int_{-\infty}^{\infty} e^{at} e^{-j\omega t} dt + \int_{0}^{\infty} e^{-at} e^{-j\omega t} dt = \int_{-\infty}^{\infty} e^{(a-j\omega)t} dt + \int_{0}^{\infty} e^{-(a+j\omega)t} dt$$

$$\int_0^{\infty} e^{-(a-j\omega)t} dt + \int_0^{\infty} e^{-(a+j\omega)t} dt = \left[\frac{e^{-(a-j\omega)t}}{-(a-j\omega)}\right]_{.0}^{\infty} + \left[\frac{e^{-(a+j\omega)t}}{-(a+j\omega)}\right]_{.0}^{\infty}$$

$$\frac{e^{-\infty} - e^{-0}}{-(a-j\omega)} + \frac{e^{-\infty} - e^{-0}}{-(a+j\omega)} = \frac{1}{a - j\omega} + \frac{1}{a + j\omega} = \frac{2a}{a^2 + \omega^2}$$

$$\therefore F(e^{-a[t]}) = \frac{2a}{a^2 + \omega^2} \text{ or } e^{-a[t]} \overset{FT}{\leftrightarrow} \frac{2a}{a^2 + \omega^2}$$

$$\therefore [x(\omega)] = \frac{2a}{a^2 + \omega^2} \text{ for all } \omega$$

And $\lfloor x(\omega) \rfloor = 0$ for allω

A Two sided exponential function and its amplitude and phase spectra are shown in figures below:

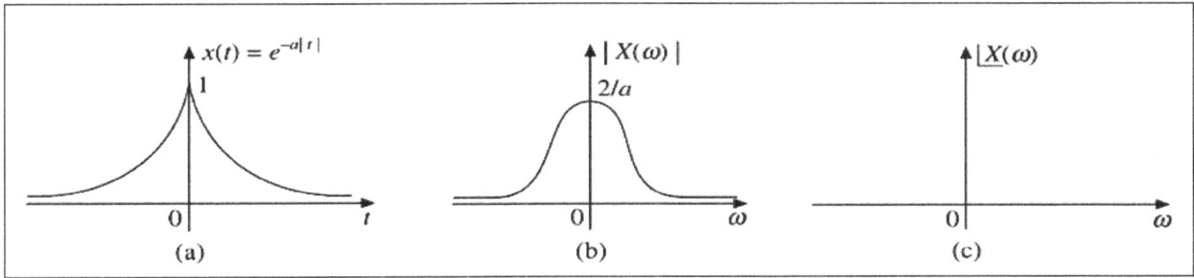

Complex Exponential Function $e^{j\omega_0 t}$

To find the Fourier Transform of complex exponential function $e^{j\omega_0 t}$, consider finding the inverse Fourier transform of $\delta(\omega - \omega_0)$. Let,

$$\omega) = \delta\ (\omega - \omega$$

$$\therefore x(t) = F^{-1}[X(\omega)] = F^{-1}[\delta\ (\omega - \omega_0)] = \frac{1}{2\pi}\int_{-\infty}^{\infty} X(\omega)\, e^{j\omega t} d\omega$$

$$= \frac{1}{2\pi}\int_{-\infty}^{\infty} \delta(\omega - \omega_0)\ e^{j\omega t} d\omega = \frac{1}{2\pi} e^{\cdot \omega_0 t}$$

$$\therefore F^{-1}[\delta(\omega - \omega_0)] = \frac{e^{j\omega_0 t}}{2\pi} \text{ or } F^{-1}[2\pi\delta(\omega - \omega_0)] = e^{j\omega_0 t}$$

$$= F[e^{j\omega_0 t}] = 2\pi\delta(\omega - \omega_0)$$

Or $e^{j\omega_0 t} \overset{FT}{\leftrightarrow} 2\pi\delta(\omega - \omega_0)$

Constant Amplitude

Let $x(t) = 1 \quad -\infty \le t \le \infty$

The waveform of a constant function is shown in below figure .Let us consider a small section of constant function, say, of duration τ.If we extend the small duration to infinity, we will get back the original function. Therefore,

$$x(t) = \underset{t \to \infty}{\text{Lt}} [\text{rect}\left(\frac{t}{\tau}\right)]$$

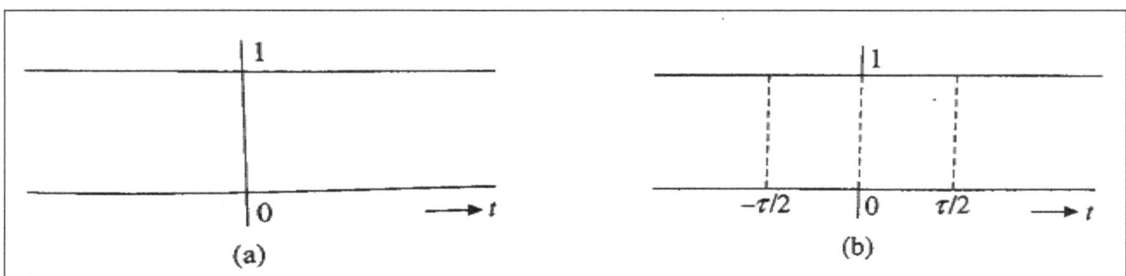

Where $\text{rect}\left(\frac{t}{\tau}\right) = \begin{cases} 1 & \text{for } \frac{-\tau}{2} \le t \le \frac{\tau}{2} \\ 0 & \text{elesewhere} \end{cases}$

By definition, the Fourier transform of $x(t)$ is:

$$x(\omega) = F[x(t)] = F\left[\underset{t\to\infty}{\text{Lt}}\ \text{rect}\left(\frac{t}{\tau}\right)\right] = \underset{t\to\infty}{\text{Lt}}\ F\left[\text{rect}\left(\frac{t}{\tau}\right)\right]$$

$$= \underset{t\to\infty}{\text{Lt}} \int_{-\tau/2}^{\tau/2} (1)e^{-j\omega t}\, dt = \underset{t\to\infty}{\text{Lt}} \left[\frac{e^{-j\omega t}}{-j\omega}\right]_{-\tau/2}^{\tau/2}$$

$$= \underset{t\to\infty}{\text{Lt}} \left[\frac{e^{-j\omega t(\tau/2)} - e^{j\omega(\tau/2)}}{-j\omega}\right] = \underset{t\to\infty}{\text{Lt}} \left\{\frac{2\sin[\omega\left(\frac{\tau}{2}\right)]}{\omega}\right\} = \underset{t\to\infty}{\text{Lt}} \left\{\hat{o}\frac{\sin[\omega\left(\frac{\tau}{2}\right)]}{\omega(\frac{\tau}{2})}\right\}$$

$$= \underset{t\to\infty}{\text{Lt}}\ \tau\ \text{sa}\left(\frac{\omega\tau}{2}\right) = 2\pi\left[\underset{t\to\infty}{\text{Lt}} \frac{\tau/2}{\pi} \text{sa}\left(\frac{\omega\tau}{2}\right)\right]$$

Using the sampling property of the delta function $= \left\{\text{i.e.}\left[\underset{t\to\infty}{\text{Lt}} \frac{\tau/2}{\pi} \text{sa}\left(\frac{\omega\tau}{2}\right)\right] = \delta(\omega)\right\}$, we get,

$$X(\omega) = F\left[\underset{t\to\infty}{\text{Lt}}\ \text{rect}\ \frac{t}{\tau}\right] = 2\pi\delta(\omega)$$

Signum Function sgn(t)

The signum function is denoted by $\text{sgn}(t)$ and is defined by:

$$\text{sgn}(t) = \begin{cases} 1 & \text{for } t > 0 \\ -1 & \text{for } t < 0 \end{cases}$$

This function is not absolutely integrable. So we cannot directly find its Fourier transform. Therefore, let us consider the function $e^{-a|t|}\,\text{sgn}(t)$ and substitute the limit a→o to obtain the above $\text{sgn}(t)$:

$$\text{Given } x(t) = \text{sgn}(t) = \underset{a\to 0}{\text{Lt}}\ e^{-a|t|}\,\text{sgn}(t) = \underset{a\to 0}{\text{Lt}}[e^{-at}u(t) - e^{-at}u(-t)$$

$$\therefore X(\omega) = F\,[\mathrm{sgn}(t)] = \int_{-\infty}^{\infty} \underset{a\to 0}{\mathrm{Lt}}[e^{-at}u(t) - e^{-at}u(-t)e^{-j\omega t}dt]$$

$$= \underset{a\to 0}{\mathrm{Lt}}\left[\int_{-\infty}^{\infty} e^{-at}e^{-j\omega t}u(t)dt - \int_{-\infty}^{\infty} e^{at}e^{-j\omega t}u(-t)dt\right]$$

$$= \underset{a\to 0}{\mathrm{Lt}}\left[\int_{0}^{\infty} e^{-(a+j\omega t)t}dt - \int_{-\infty}^{0} e^{(a-j\omega)t}dt\right] = \underset{a\to 0}{\mathrm{Lt}}\left[\int_{0}^{\infty} e^{-(a+j\omega)t}dt - \int_{0}^{\infty} e^{-(a-j\omega)t}dt\right]$$

$$F[\mathrm{sgn}(t)] = \frac{2}{j\omega}$$

$$\mathrm{sgn}(t) \overset{FT}{\leftrightarrow} \frac{2}{j\omega}$$

$$\therefore \quad |X(\omega)| = \frac{2}{\omega} \text{ and } \left\lfloor X(\omega) = \frac{\pi}{2} f\,or\,\omega < 0 \; and -\frac{\pi}{2} f\,or\,\omega > 0 \right\rfloor$$

Figure below shows the signum function and its magnitude and phase spectra.

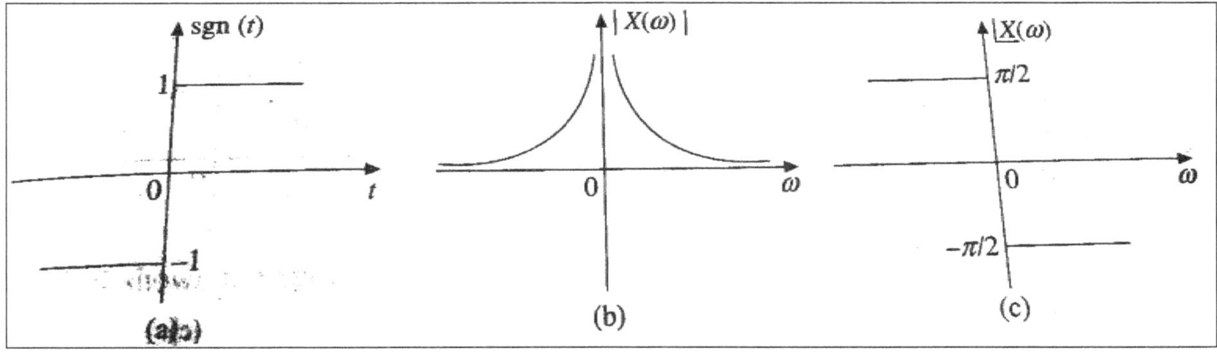

Unit step function $u(t)$.

The unit step function is defined by:

$$u(t) = \begin{cases} 1 & \text{for } t \geq 0 \\ 0 & \text{for } t < 0 \end{cases}$$

since the unit step function is not absolutely integrable, we cannot directly find its Fourier transform. So express the unit step function in terms of signum function as:

$$u(t) = \frac{1}{2} + \frac{1}{2}\mathrm{sgn}\,(t)$$

$$x(t) = u(t) = \frac{1}{2}[1 + \mathrm{sgn}\,(t)]$$

$$X(\omega) = F[u(t)] = F\left\{\frac{1}{2}[1+\text{sgn}(t)]\right\}$$

$$= \frac{1}{2}\{F[1] + F[\text{sgn}(t)]\}$$

We know that $F[1] = 2\pi\delta(\omega)$ and $F[\text{sgn}(t)] = \frac{2}{j\omega}$,

$$u(t) \overset{FT}{\leftrightarrow} \pi\delta(\omega) + \frac{1}{j\omega}$$

$\therefore |X(\omega)| = \infty$ at $\omega = 0$ and is equal to 0 at $\omega = -\infty$ and $\omega = \infty$. Figure shows the unit step function and its spectrum.

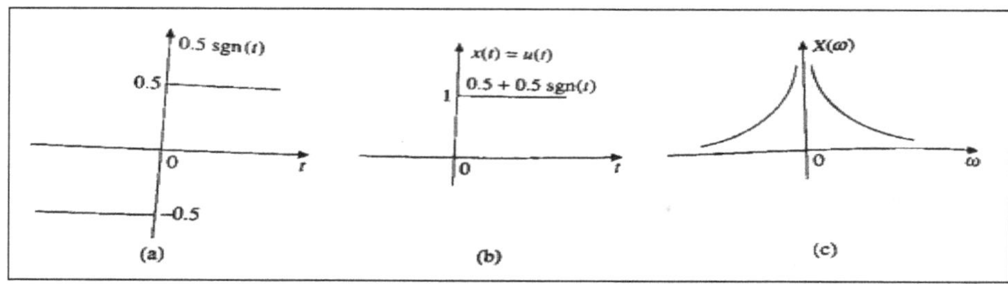

Rectangular Pulse (Gate Pulse) $\Pi\left(\frac{t}{\tau}\right)$ or rect$\left(\frac{t}{\tau}\right)$

Consider a rectangular pulse as shown in below figure. This is called a unit gate function and is defined as:

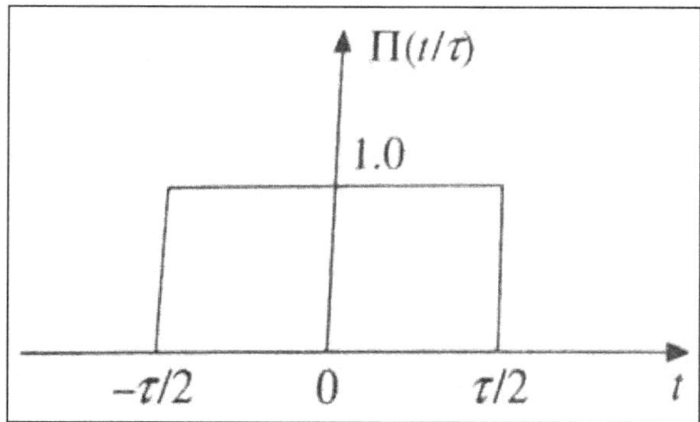

$$x(t) = \text{rect}\left(\frac{t}{\tau}\right) = \Pi\left(\frac{t}{\tau}\right) = \begin{cases} 1 & \text{for } |t| < \frac{T}{2} \\ 0 & \text{otherwise} \end{cases}$$

Then,

$$X(\omega) = F[x(t)] = F\left[\Pi\left(\frac{t}{\tau}\right)\right] = \int_{-\infty}^{\infty} \Pi\left(\frac{t}{\tau}\right) e^{-j\omega t} dt$$

$$= \int_{-\tau/2}^{\tau/2} (1)e^{-j\omega t}dt = \left[\frac{e^{-j\omega t}}{-j\omega}\right]_{\tau/2}^{\tau/2} = \frac{e^{-j\omega t(\tau/2)} - e^{j\omega(\tau/2)}}{-j\omega}$$

$$= \frac{\tau}{\omega(-\tau/2)}\left[\frac{e^{j\omega(\tau/2)} - e^{-j\omega(\tau/2)}}{2j}\right] = \tau\left[\frac{\sin\omega(\tau/2)}{\omega(\tau/2)}\right]$$

$$= \tau \operatorname{sinc} \omega(\tau/2)$$

$$\therefore F\left[\Pi\left(\frac{t}{\tau}\right)\right] = \tau \operatorname{sinc}\ \omega(\tau/2), \text{ that is}$$

$$\operatorname{rect}\left(\frac{t}{\tau}\right) = \Pi\left(\frac{t}{\tau}\right) \overset{FT}{\leftrightarrow} \tau \operatorname{sinc}\ \omega(\tau/2)$$

Figure shows the spectra of the gate function.

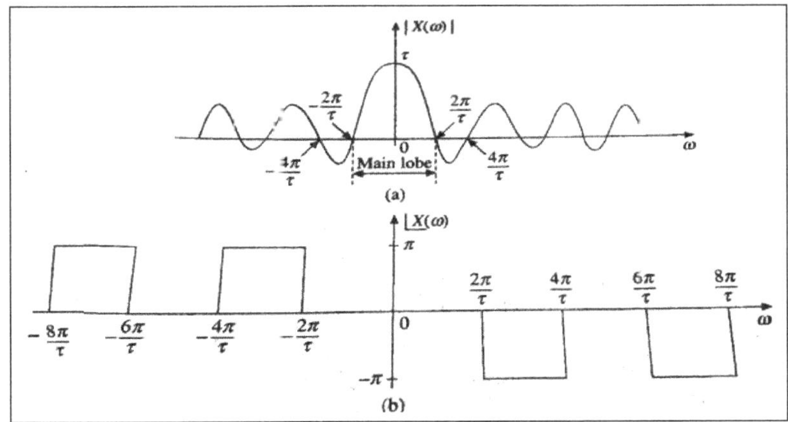

The amplitude spectrum is obtained as follows:

At $\omega = 0$, sinc $(\omega\tau/2) = 1$. Therefore, $|X(\omega)|$ at ù $= 0$ is equal to τ.At$(\omega\tau/2) = \pm n\pi$, at

$$\omega = \pm\frac{2n\pi}{\pi}, n = 1,2,\ldots\ldots \operatorname{sinc}\ (\omega\tau/2) = 0$$

The phase spectrum is:

$$\lfloor X(\omega) \rfloor = 0 \quad \text{if sinc } (\omega\tau/2) > 0$$
$$= \pm\pi \text{ if sinc } (\omega\tau/2) < 0$$

The amplitude response between the first two zero crossings is known as main lobe and the portions of the response for $\omega < -\left(\frac{2\pi}{\tau}\right)$ and $\omega > \left(\frac{2\pi}{\tau}\right)$ are known as side lobes. From the amplitude spectrum, we can find that majority of the energy of the signal is contained in the main lobe.

The first zero crossing occurs at $\omega = \left(\frac{2\pi}{\tau}\right)$ or at $f = \frac{1}{\tau}$ Hz. As the width of the rectangular pulse is made longer, the main lobe becomes narrower. The phase spectrum is odd function of ω. If the amplitude spectrum is positive, then phase is zero, and if the amplitude spectrum is negative, then the phase is $-\pi$ *or* π.

Triangular Pulse $\Delta\left(\frac{t}{\tau}\right)$

Consider the triangular pulse as shown in below figure. It is defined as:

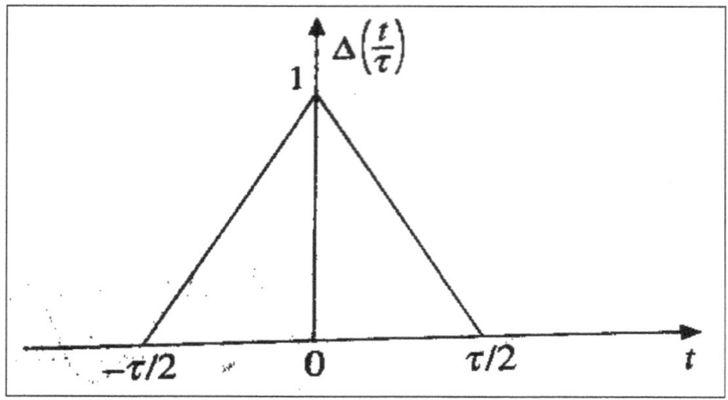

$$x(t) = \Delta\left(\frac{t}{\tau}\right) = \begin{cases} \frac{1}{\tau/2}\left(t+\frac{\tau}{2}\right) = \left(1+2\frac{t}{\tau}\right) & \text{for } -\frac{\tau}{2} < t < 0 \\ \frac{1}{\tau/2}\left(t-\frac{\tau}{2}\right) = \left(1-2\frac{t}{\tau}\right) & \text{for } 0 < t < \frac{\tau}{2} \\ & \text{elsewhere} \end{cases}$$

$$x(t) = \Delta\left(\frac{t}{\tau}\right) = \begin{cases} 1-\frac{2|t|}{\tau} & \text{for } |t| < \frac{T}{2} \\ 0 & \text{otherwise} \end{cases}$$

$$\text{Then } X(\omega) = F[x(t)] = F\left[\Delta\left(\frac{t}{\hat{o}}\right)\right] = \int_{-\infty}^{\infty} \Delta\left(\frac{t}{\hat{o}}\right) e^{-j\omega t} dt$$

$$= \int_{-\tau/2}^{0}\left(1+\frac{2t}{\hat{o}}\right)e^{-j\omega t}dt + \int_{0}^{\tau/2}\left(1-\frac{2t}{\tau}\right)e^{-j\omega t}dt$$

$$= \int_{0}^{\tau/2}\left(1-\frac{2t}{\tau}\right)e^{j\omega t}dt + \int_{0}^{\tau/2}\left(1-\frac{2t}{\tau}\right)e^{-j\omega t}dt$$

$$= \int_{0}^{\tau/2} e^{j\omega t}\, dt - \int_{0}^{\tau/2}\left(\frac{2t}{\tau}\right)e^{j\omega t}\, dt + \int_{0}^{\tau/2} e^{-j\omega t}\, dt - \int_{0}^{\tau/2}\left(\frac{2t}{\tau}\right)e^{-j\omega t}dt$$

$$= \int_{0}^{\tau/2} e^{j\omega t}\, dt - \int_{0}^{\tau/2}\left(\frac{2t}{\tau}\right)e^{j\omega t}\, dt + \int_{0}^{\tau/2} e^{-j\omega t}\, dt - \int_{0}^{\tau/2}\left(\frac{2t}{\tau}\right)e^{-j\omega t}dt$$

$$= \int_0^{\tau/2} 2\cos\omega t\,dt - \frac{2}{\tau}\int_0^{\tau/2} 2t\cos\omega t\,dt$$

$$= 2\left[\frac{\sin\omega t}{\omega}\right]_0^{\tau/2} - \frac{4}{\tau}\left[\left[t\frac{\sin\omega t}{\omega}\right]_0^{\tau/2} + \left[\frac{\cos\omega t}{\omega^2}\right]_0^{\tau/2}\right]$$

$$= \frac{2}{\omega}\left[\sin\omega\frac{\tau}{2}\right] - \frac{4}{\omega\tau}\left[\frac{\tau}{2}\sin\frac{\omega\tau}{2}\right] - \frac{4}{\omega^2\tau}\left[\cos\frac{\omega\tau}{2} - 1\right]$$

$$= \frac{4}{\omega^2\tau}\left[1 - \cos\frac{\omega\tau}{2}\right] = \frac{4}{\omega^2\tau}\left[2\sin^2\frac{\omega\tau}{2}\right]$$

$$= \frac{8}{\omega^2\tau}\left(\frac{\omega\tau}{4}\right)^2 \frac{\sin^2\left(\frac{\omega\tau}{4}\right)}{\left(\frac{\omega\tau}{4}\right)} = \frac{\tau}{2}\text{sinc}^2\left(\frac{\omega\tau}{4}\right)$$

$$F\left[\Delta\left(\frac{t}{\tau}\right)\right] = \frac{\tau}{2}\text{sinc}^2\left(\frac{\omega\tau}{4}\right)$$

$$\text{Or} \quad \Delta\left(\frac{t}{\tau}\right) \overset{FT}{\leftrightarrow} \frac{\tau}{2}\text{sinc}^2\left(\frac{\omega\tau}{4}\right)$$

Figure shows the amplitude spectrum of a triangular pulse.

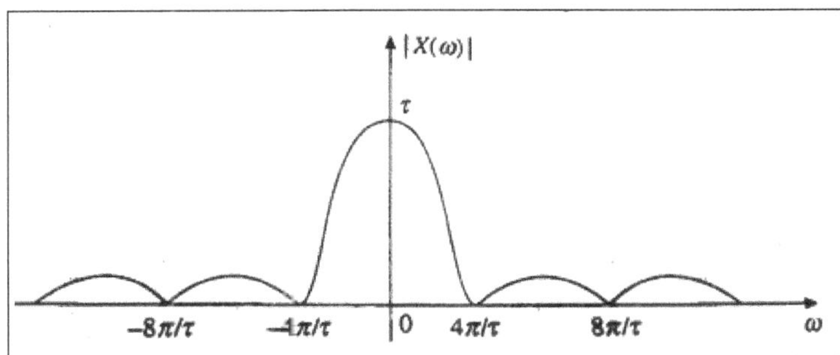

Cosine Wave $(\cos \omega_0 t)$

Given $x(t) = \cos \omega_0 t$

Then $X(\omega) = F[x(t)] = F[\cos\omega_0 t] = F\left[\frac{1}{2}(e^{j\omega_0 t} + e^{-j\omega_0 t})\right]$

$$= \frac{1}{2}[F(e^{j\omega_0 t}) + F(e^{-j\omega_0 t})] = \frac{1}{2}[2\pi\delta(\omega - \omega_o) + 2\pi\delta(\omega + \omega_o)]$$

$$= \pi[\delta(\omega - \omega_o) + \delta(\omega - \omega_o)]$$

$$\therefore F[\cos\,\omega_0 t] = \pi[\delta(\omega - \omega_o) + \delta(\omega + \omega_o)] \text{ or } \cos\,\omega_0 t \overset{FT}{\leftrightarrow} \pi[\delta(\omega - \omega_o) + \delta(\omega + \omega_o)]$$

Below figure shows the cosine wave and its amplitude and phase spectra.

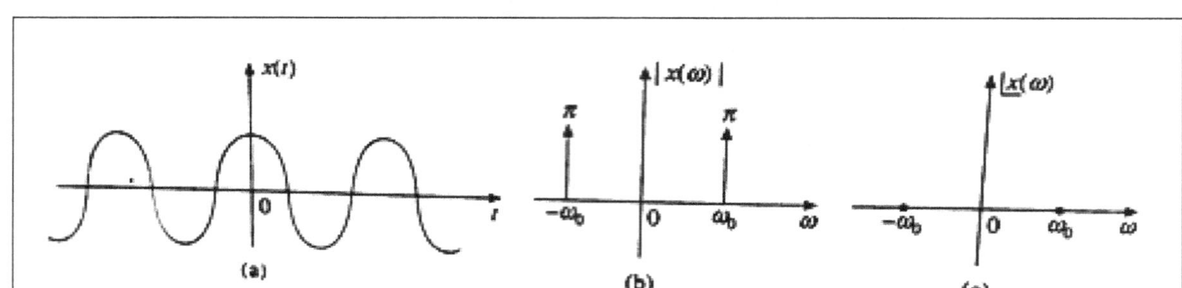

Sine Wave ($\sin\omega_0 t$)

Given $x(t) = \sin \omega_0 t$

$$\text{Then } X(\omega) = F[x(t)] = F[\sin \omega_0 t] = F\left[\frac{1}{2j}(e^{j\omega_0 t} + e^{-j\omega_0 t})\right]$$

$$= \frac{1}{2j}[F(e^{j\omega_0 t}) + F(e^{-j\omega_0 t})] = \frac{1}{2}[2\pi\delta(\omega - \omega_o) + 2\pi\delta(\omega + \omega_o)]$$

$$= -j\pi[\delta(\omega - \omega_o) + \delta(\omega + \omega_o)]$$

$$\therefore F[\cos \omega_0 t] = -j\pi[\delta(\omega - \omega_o) - \delta(\omega + \omega_o)] \text{ or } \cos \omega_0 t \overset{FT}{\leftrightarrow} j\pi[\delta(\omega - \omega_o) + \delta(\omega + \omega_o)]$$

Below figure shows the sine wave and its amplitude and phase spectra.

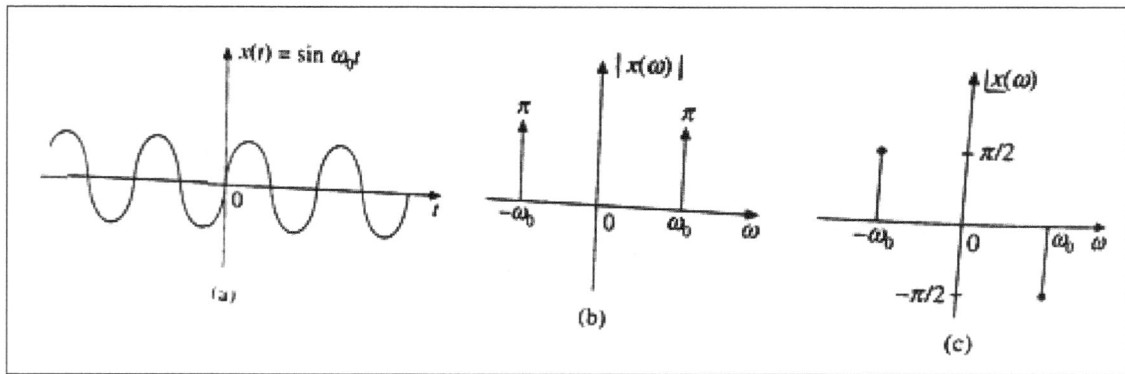

Fourier Transform of a Periodic Signal

The periodic functions can be analysed using Fourier series and that non-periodic function can be analysed using Fourier transform. But we can find the Fourier transform of a periodic function also. This means that the Fourier transform can be used as a universal mathematical tool in the analysis of both non-periodic and periodic waveforms over the entire interval. Fourier transform of periodic functions may be found using the concept of impulse function.

We know that using Fourier series, any periodic signal can be represented as a sum of complex exponentials. Therefore, we can represent a periodic signal using the Fourier integral. Let us consider a periodic signal x(t) with period T. Then, we can express x(t) in terms of exponential Fourier series as:

$$x(t) = \sum_{n=-\infty}^{\infty} C_n e^{jn} \omega_0 t$$

The Fourier transform of x(t) is:

$$X(\omega) = F[x(t)] = F\left[\sum_{n=\infty}^{\infty} C_n e^{jn\omega_0 t}\right]$$
$$= \sum_{n=\infty}^{\infty} C_n F[e^{jn\omega_0 t}]$$

Using the frequency shifting theorem, we have:

$$F[1e^{jn\omega_0 t}] = F[1]\Big|_{\omega=\omega-n\omega_0} = s\pi\delta(\omega - n\omega_0)$$
$$X(\omega) = 2\pi \sum_{n=\infty}^{\infty} C_n \delta(\omega - n\omega_0)$$

where $C_n s$ are the Fourier coefficients associated with x(t) and are given by:

$$C_n = \frac{1}{T}\int_{-T/2}^{T/2} x(t)e^{-jn\omega_0 t}\,dt$$

Thus, the Fourier transform of a periodic function consists of a train of equally spaced impulses. These impulses are located at the harmonic frequencies of the signal and the strength of each impulse is given as $2\pi C_n$.

Solved Problems:

Problem : Find the Fourier transform of the signals $e^{3t}(t)$.

Solution:

Given,

$$x(t)=e^{3t}u(t)$$

The given signal is not absolutely integrable. That is $\int_{-\infty}^{\infty} e^{3t}u(t) = \infty$. Therefore, Fourier transform of $x(t)=e^{3t}u(t)$ does not exist.

Problem: Find the Fourier transform of the signals $\cos\omega_0 t\,u(t)$.

Solution:

$$\text{Given} \quad x(t) = \cos\omega_0 t\,u(t)$$

$$\text{i.e.} \qquad = \frac{e^{j\omega_0 t} + e^{-j\omega_0 t}}{2} u(t)$$

$$\therefore \quad X(\omega) = F[\cos\omega_0 t\, u(t)] = \int_{-\infty}^{\infty} \frac{e^{j\omega_0 t} + e^{-j\omega_0 t}}{2} u(t) e^{-j\omega t} dt$$

$$= \frac{1}{2}\left[\int_0^{\infty} e^{-j(\omega-\omega_0)t} dt + \int_0^{\infty} e^{-j(\omega+\omega_0)t} dt\right]$$

$$= \frac{1}{2}\left[\frac{e^{-j(\omega-\omega_0)t}}{-j(\omega-\omega_0)} + \frac{e^{-j(\omega+\omega_0)t}}{-j(\omega+\omega_0)}\right]_0^{\infty}$$

$$= \frac{1}{2}\left[\frac{-e^0}{-j(\omega-\omega_0)} + \frac{-e^0}{-j(\omega+\omega_0)}\right]$$

With impulses of strength π at $\omega = \omega_0$ and $\omega = -\omega_0$.

$$\therefore \quad X(\omega) = \frac{1}{2}\left[\frac{1}{j(\omega-\omega_0)} + \frac{1}{j(\omega+\omega_0)}\right] + \pi\delta(\omega-\omega_0) + \pi\delta(\omega+\omega_0)]$$

$$= \frac{1}{2}\left[\frac{j2\omega}{(j\omega)^2 - \omega_0^2} + \pi\delta(\omega-\omega_0) + \pi\delta(\omega+\omega_0)\right]$$

$$= \frac{j\omega}{(j\omega)^2 - \omega_0^2} + \frac{1}{2}[\pi\delta(\omega-\omega_0) + \pi\delta(\omega+\omega_0)]$$

Problem: Find the Fourier transform of the signals $\sin\omega_0 t\ u(t)$.

Solution:

Given $x(t) = \sin\omega_o t\, u(t)$

i.e. $= \dfrac{e^{j\omega_o t} - e^{-j\omega_o t}}{2j} u(t)$

$$\therefore \quad X(\omega) = F[\sin\omega_0 t\, u(t)] = \int_{-\infty}^{\infty} \frac{e^{j\omega_0 t} - e^{-j\omega_0 t}}{2} u(t) e^{-j\omega t} dt$$

$$= \frac{1}{2j}\left[\int_0^{\infty} e^{-j(\omega-\omega_0)t} dt - \int_0^{\infty} e^{-j(\omega+\omega_0)t} dt\right]$$

$$= \frac{1}{2j}\left[\frac{e^{-j(\omega-\omega_0)t}}{-j(\omega-\omega_0)} - \frac{e^{-j(\omega+\omega_0)t}}{-j(\omega+\omega_0)}\right]_0^{\infty}$$

$$= \frac{1}{2j}\left[\frac{-e^0}{-j(\omega-\omega_0)} - \frac{-e^0}{-j(\omega+\omega_0)}\right]$$

With impulses of strength π at $\omega = \omega_0$ and $\omega = -\omega_0$.

$$\therefore \quad X(\omega) = \frac{1}{2j}\left[\frac{1}{j(\omega-\omega_0)} - \frac{1}{j(\omega+\omega_0)} + \pi\delta(\omega-\omega_0) - \pi\delta(\omega+\omega_0)\right]$$

$$=\frac{1}{2j}\left[\frac{j2\omega_0}{(j\omega)^2+\omega_0^2}+\pi\delta(\omega-\omega_0)\ -\ \pi\delta(\omega+\omega_0)\right]$$

$$=\frac{\omega_0}{(j\omega)^2+\omega_0^2}-j\frac{\pi}{2}[\delta(\omega-\omega_0)+\delta(\omega+\omega_0)]$$

Continuous Time Fourier Transform

The Fourier expansion coefficient $X[k]$ (a_k in OWN) of a periodic signal $x_T(t)=x_T(t+T)$ is:

$$X[k]=\frac{1}{T}\int_T x_T(t)e^{-jk\omega_0 t}dt \qquad (k=0,\pm1,\pm2,\cdots)$$

and the Fourier expansion of the signal is:

$$x_T(t)=\sum_{k=-\infty}^{\infty}X[k]e^{jk\omega_0 t}$$

which can also be written as:

$$x_T(t)=\frac{1}{T}\sum_{k=-\infty}^{\infty}(TX[k])e^{jk\omega_0 t}=\frac{\omega_0}{2\pi}\sum_{k=-\infty}^{\infty}X(k\omega_0)e^{jk\omega_0 t} \qquad (a)$$

where $X(k\omega_0)$ is defined as:

$$X(k\omega_0)\triangleq T\ X[k]=\int_T x_T(t)e^{-jk\omega_0 t}dt \qquad (b)$$

When the period of $x_T(t)$ approaches infinity $T\to\infty$, the periodic signal $x_T(t)$ becomes a non-periodic signal $x(t)$ and the following will result:

Interval between two neighboring frequency components becomes zero:

$$T\to\infty\Rightarrow\omega_0=2\pi/T\to 0$$

Discrete frequency becomes continuous frequency:

$$k\omega_0\,|_{\omega_0\to 0}\Rightarrow\omega$$

Summation of the Fourier expansion in equation (a) becomes an integral:

$$x(t)\triangleq\lim_{T\to\infty}x_T(t)=\lim_{\omega_0\to 0}\frac{1}{2\pi}\sum_{k=-8}^{\infty}X(k\omega_0)e^{jk\omega_0 t}\omega_0=\frac{1}{2\pi}\int_{-\infty}^{\infty}X(\omega)e^{j\omega t}d\omega$$

The second equal sign is due to the general fact:

$$\lim_{\Delta x \to 0} \sum_{k=-\infty}^{\infty} f(k\Delta x)\Delta x = \int_{-\infty}^{\infty} f(x)dx$$

Time integral over T in equation becomes over the entire time axis:

$$X(\omega) \overset{\Delta}{=} \lim_{T\to\infty} X(k\omega_0) = \lim_{T\to\infty} \int_T x_T(t)e^{-jk\omega_0 t}dt = \int_{-\infty}^{\infty} x(t)e^{-j\omega t}dt$$

In summary, when the signal is non-periodic $x(t) = \lim_{T\to\infty} x_T(t)$, the Fourier expansion becomes Fourier transform. The forward transform (analysis) is:

$$X(\omega) = \int_{-\infty}^{\infty} x(t)e^{-j\omega t}dt \text{ or } X(f) = \int_{-\infty}^{\infty} x(t)e^{-j2\pi f t}dt$$

and the inverse transform (synthesis) is:

$$x(t) = \frac{1}{2\pi}\int_{-\infty}^{\infty} X(\omega)e^{j\omega t}d\omega = \int_{-\infty}^{\infty} X(f)e^{j2\pi f t}df$$

Note that $X(\omega)$ is denoted by $X(j\omega)$ in own.

Comparing Fourier coefficient of a periodic signal $x_T(t)$ with Fourier spectrum of a non-periodic signal $x(t)$:

$$X[k] = \frac{1}{T}\int_T x_T(t)e^{-jk\omega_0 t}dt, \qquad X(\omega) = \int_{-\infty}^{\infty} x(t)e^{-j\omega t}dt$$

we see that the dimension of $X(\omega)$ is different from that of $X[k]$:

If $|X[k]|^2$ represents the energy contained in the kth frequency component of a periodic signal $x_T(t)$, then $|X(\omega)|^2$ represents the energy density of a non-periodic signal $x(t)$ distributed along the frequency axis. We can only speak of the energy contained in a particular frequency band $\omega_1 < \omega < \omega_2$:

$$\text{Energy contained in band } \omega_1 < \omega_2 = \int_{\omega_1}^{\omega_2} |X(\omega)|^2 \, d\omega$$

The spectrum of a time signal can be denoted by $X(\omega)$ or $X(f)$ to emphasize the fact that the spectrum represents how the energy contained in the signal is distributed as a function of frequency ω or f. Moreover, if $X(f)$ is used, the factor $1/2\pi$ in front of the inverse transform is dropped so that the transform pair takes a more symmetric form. On the other hand, as Fourier transform can be considered as a special case of Laplace transform when the real part σ of the complex argument $s = \sigma + j\omega = j\omega$ is zero:

$$X(s)|_{s=j\omega} = \int_{-\infty}^{\infty} x(t)e^{-st}dt\Big|_{s=j\omega} = \int_{-\infty}^{\infty} x(t)e^{-j\omega t}dt = X(j\omega)$$

It is also natural to denote the spectrum of $x(t)$ by $X(j\omega)$ (in OWN).

Example:

Consider the unit impulse function:

$$x(t) = \delta(t)$$

$$X(j\omega) = \int_{-\infty}^{\infty} \delta(t) e^{-j\omega t} dt = 1$$

Example:

If the spectrum of a signal $x(t)$ is a delta function in frequency domain $X(j\omega) = 2\pi\,\delta(\omega)$, the signal can be found to be:

$$x(t) = F^{-1}[X(j\omega)] = \frac{1}{2\pi}\int_{-\infty}^{\infty} 2\pi\,\delta(\omega) e^{j\omega t} d\omega = e^0 = 1$$

i.e.,

$$F[x(t)] = \int_{-\infty}^{\infty} e^{-j\omega t} dt = 2\pi\,\delta(\omega)$$

Example:

The spectrum is:

$$X(j\omega) = \int_{-a}^{a} e^{-j\omega t} dt = \frac{1}{-j\omega} e^{-j\omega t}\Big|_{-a}^{a} = \frac{2}{\omega} sin(a\omega)$$

This is the sinc function with a parameter a, as shown in the figure.

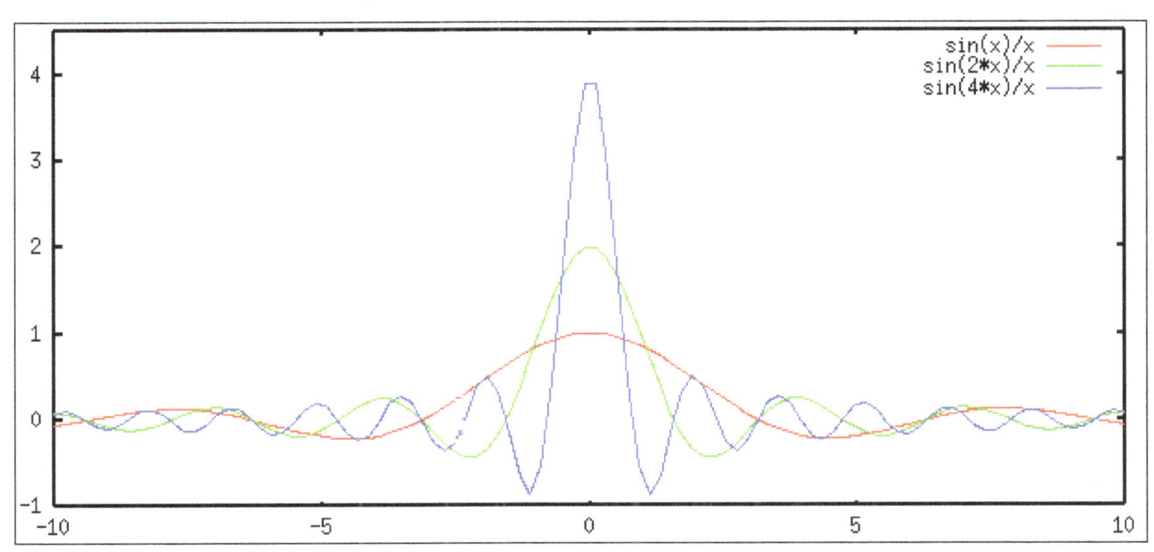

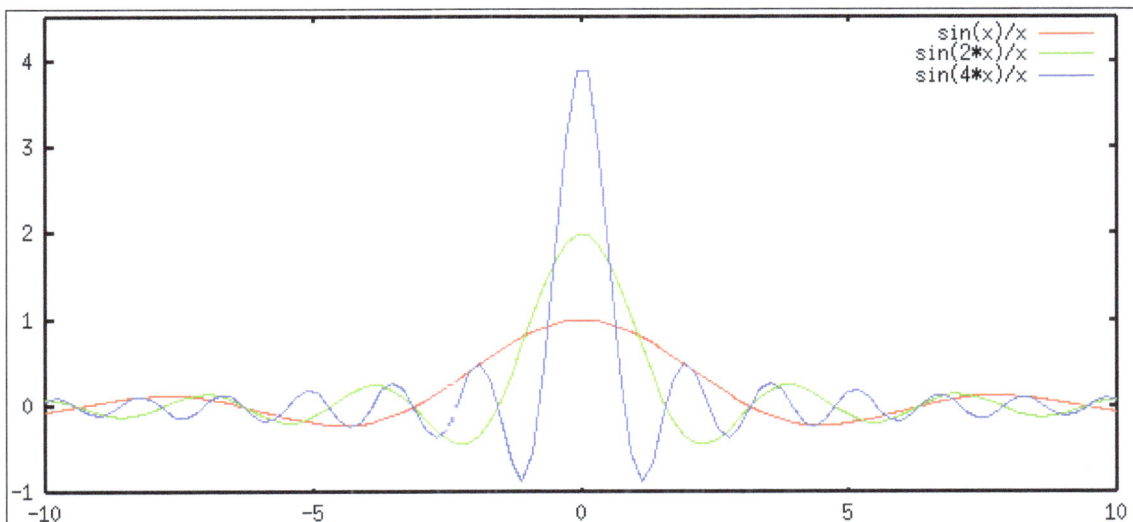

Note that the height of the main peak is $2a$ and it gets taller and narrower as a gets larger. Also note:

$$\int_{-\infty}^{\infty} X(j\omega)d\omega = 2\int_{-\infty}^{\infty} \frac{sin(a\omega)}{\omega} d\omega = 2\pi$$

When a approaches infinity, $x(t) = 1$ for all t, and the spectrum becomes:

$$X(j\omega) = \int_{-\infty}^{\infty} e^{-j\omega t} dt = \lim_{a-\infty} [\frac{2}{\omega} \sin(a\omega)] = 2\pi\delta(\omega) = \delta(f)$$

Recall that the Fourier coefficient of $x(t) = 1$ is:

$$X[k] = \delta[k] = \begin{cases} 1 & k = 0 \\ 0 & \text{else} \end{cases}$$

which represents the energy contained in the signal at $k = 0$ (DC component at zero frequency), and the spectrum $X(j\omega) = X[k]/\omega$ is the energy density or distribution which is infinity at zero frequency.

The integral in the above transform is an important formula to be used frequently later:

$$\int_{-\infty}^{\infty} e^{-j\omega t} dt = 2\pi\delta(\omega) \quad or \quad \int_{-\infty}^{\infty} e^{-j2\pi ft} dt = \delta(f)$$

which can also be written as:

$$\int_{-\infty}^{\infty} e^{-j\omega t} dt = \int_{-\infty}^{\infty} [\cos(\omega t) + j\sin(-j\omega t)]dt = \int_{-\infty}^{\infty} \cos(\omega t)dt = \delta(f)$$

Switching t and f in the equation above, we also have:

$$\int_{-\infty}^{\infty} e^{-j2\pi ft} df = \cos(2\pi ft)df = \delta(t)$$

representing a superposition of an infinite number of cosine functions of all frequencies, which cancel each other any where along the time axis except at $t = 0$ where they add up to infinity, an impulse.

Example:

$$x(t) = cos(\omega_0 t) = \frac{1}{2}[e^{j\omega_0 t} + e^{-j\omega_0 t}]$$

The spectrum of the cosine function is:

$$X(j\omega): \quad \int_{-\infty}^{\infty} x(t)e^{-j\omega t} dt = \frac{1}{2}\int_{-\infty}^{\infty} [e^{j\omega_0 t} + e^{-j\omega_0 t}]e^{-j\omega_0 t} dt$$

$$: \quad \pi[\delta(\omega - \omega_0) + \delta(\omega + \omega_0)]$$

The spectrum of the sine function:

$$x(t) = sin(\omega_0 t) = \frac{1}{2j}[e^{j\omega_0 t} - e^{-j\omega_0 t}]$$

can be similarly obtained to be:

$$X(j\omega) = \frac{\pi}{j}[\delta(\omega - \omega_0) - \delta(\omega + \omega_0)] = -j\pi[\delta(\omega - \omega_0) - \delta(\omega + \omega_0)]$$

Again, these spectra represent the energy density distribution of the sinusoids, while the corresponding Fourier coefficients:

$$X[k] = F[cos(\omega_0 t)] = \frac{1}{2}[\delta[k-1] + \delta[k+1]]$$

And

$$X[k] = F[\sin(\omega_0 t)] = \frac{1}{2j}[\delta[k-1] - \delta[k+1]]$$

Represent the energy contained at frequency $\omega = \omega_0$.

Properties of Continuous-time Fourier Transform

Linearity

The combined addition and scalar multiplication properties in the table above demonstrate the basic property of linearity. What you should see is that if one takes the Fourier transform of a linear combination of signals then it will be the same as the linear combination of the Fourier transforms of each of the individual signals. This is crucial when using a table of transforms to find the transform of a more complicated signal.

Example :

We will begin with the following signal:

$$z(t) = af_1(t) + bf_2(t)$$

Now, after we take the Fourier transform, shown in the equation below, notice that the linear combination of the terms is unaffected by the transform.

$$Z(\omega) = aF_1(\omega) + bF_2(\omega)$$

Symmetry

Symmetry is a property that can make life quite easy when solving problems involving Fourier transforms. Basically what this property says is that since a rectangular function in time is a sinc function in frequency, then a sinc function in time will be a rectangular function in frequency. This is a direct result of the similarity between the forward CTFT and the inverse CTFT. The only difference is the scaling by 2π and a frequency reversal.

Time Scaling

This property deals with the effect on the frequency-domain representation of a signal if the time variable is altered. The most important concept to understand for the time scaling property is that signals that are narrow in time will be broad in frequency and vice versa. The simplest example of this is a delta function, a unit pulse with a very small duration, in time that becomes an infinite-length constant function in frequency.

The table above shows this idea for the general transformation from the time-domain to the frequency-domain of a signal. You should be able to easily notice that these equations show the relationship mentioned previously: if the time variable is increased then the frequency range will be decreased.

Time Shifting

Time shifting shows that a shift in time is equivalent to a linear phase shift in frequency. Since the frequency content depends only on the shape of a signal, which is unchanged in a time shift, then only the phase spectrum will be altered. This property is proven below:

Example:

We will begin by letting $z(t) = f(t-\tau)$. Now let us take the Fourier transform with the previous expression substituted in for $z(t)$.

$$z(\omega) = \int_{-\infty}^{\infty} f(t-\tau)e^{-(i\omega t)}dt$$

Now let us make a simple change of variables, where $\sigma = t - \tau$. Through the calculations below, you can see that only the variable in the exponential are altered thus only changing the phase in the frequency domain.

$$\begin{aligned} z(\omega) &= \int_{-\infty}^{\infty} f(\sigma)e^{-(i\omega(\sigma+\tau)t)}d\tau \\ &= e^{-(i\omega\tau)}\int_{-\infty}^{\infty} f(\sigma)e^{-(i\omega\sigma)}d\sigma \\ &= e^{-(i\omega\tau)}F(\omega) \end{aligned}$$

Convolution

Convolution is one of the big reasons for converting signals to the frequency domain, since convolution in time becomes multiplication in frequency. This property is also another excellent example of symmetry between time and frequency. It also shows that there may be little to gain by changing to the frequency domain when multiplication in time is involved.

We will introduce the convolution integral here, but if you have not seen this before or need to refresh your memory, then look at the continuous-time convolution module for a more in depth explanation and derivation.

$$\begin{aligned} y(t) &= (f_1(t), f_2(t)) \\ &= \int_{-\infty}^{\infty} f_1(\tau) f_2(t-\tau) d\tau \end{aligned}$$

Time Differentiation

Since LTI systems can be represented in terms of differential equations, it is apparent with this property that converting to the frequency domain may allow us to convert these complicated differential equations to simpler equations involving multiplication and addition. This is often looked at in more detail during the study of the Laplace Transform.

Parseval's Relation

$$\int_{-\infty}^{\infty} (|f(t)|)^2 dt = \int_{-\infty}^{\infty} (|F(\omega)|)^2 df$$

Parseval's relation tells us that the energy of a signal is equal to the energy of its Fourier transform.

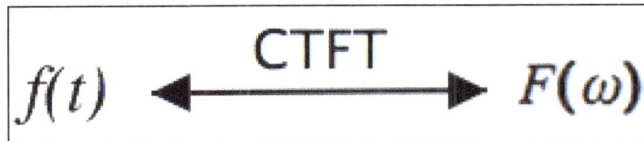

$$f(t) \xleftrightarrow{\text{CTFT}} F(\omega)$$

Modulation (Frequency Shift)

Modulation is absolutely imperative to communications applications. Being able to shift a signal to a different frequency, allows us to take advantage of different parts of the electromagnetic spectrum is what allows us to transmit television, radio and other applications through the same space without significant interference.

The proof of the frequency shift property is very similar to that of the time shift; however, here we would use the inverse Fourier transform in place of the Fourier transform.

Since we went through the steps in the previous, time-shift proof, below we will just show the initial and final step to this proof:

$$z(t) =; \frac{1}{2\pi} \int_{-\infty}^{\infty} F(\omega - \varphi) e^{i\omega t} d\omega$$

Now we would simply reduce this equation through another change of variables and simplify the terms. Then we will prove the property expressed in the table above:

$$z(t) = f(t) e^{i\varphi t}$$

References

- Fourier-series, signals-and-systems: tutorialspoint.com, Retrieved 25 April, 2019
- Bracewell, R. N. (2000), The Fourier Transform and Its Applications (3rd ed.), Boston: McGraw-Hill, ISBN 978-0-07-116043-8
- Fourier-series-properties, signals-and-systems: tutorialspoint.com, Retrieved 16 June, 2019
- Müller, Meinard (2015), The Fourier Transform in a Nutshell. (PDF), In Fundamentals of Music Processing, Section 2.1, pages 40-56: Springer, doi:10.1007/978-3-319-21945-5, ISBN 978-3-319-21944-8
- "Signals and Systems, Analysis using Transform methods and MATLAB", Second edition,McGraw-Hill Education,2011

5
Laplace Transform

A transformation which is used to convert an operation of a real variable (t) into complex variable (s) is termed as laplace transform. It is used for analysing and developing circuits such as filters. This chapter delves into various concepts such as laplace transform properties, region of convergence, the laplace transform of a function, existence of laplace transform, etc. which will provide in-depth knowledge of the subject.

The Laplace transform is an integral transform perhaps second only to the Fourier transform in its utility in solving physical problems. The Laplace transform is particularly useful in solving linear ordinary differential equations such as those arising in the analysis of electronic circuits.

The (unilateral) Laplace transform L is defined by:

$$\mathcal{L}_t[f(t)](s) = \int_0^\infty f(t)e^{-st}\,dt,$$

where $f(t)$ is defined for $t \geq 0$. The unilateral Laplace transform is almost always what is meant by "the" Laplace transform, although a bilateral Laplace transform is sometimes also defined as:

$$\mathcal{L}_t^{(2)}[f(t)](s) = \int_{-\infty}^{\infty} f(t)e^{-st}\,dt$$

The unilateral Laplace transform $\mathcal{L}_t[f(t)](s)$ is implemented in the Wolfram Language as Laplace Transform $[f[t], t, s]$ and the inverse Laplace transform as Inverse Radon Transform.

The inverse Laplace transform is known as the Bromwich integral, sometimes known as the Fourier-Mellin integral.

A table of several important one-sided Laplace transforms is given below.

f	$\mathcal{L}_t[f(t)](s)$	conditions
1	$\frac{1}{s}$	

t	$\frac{1}{s^2}$	
t^n	$\frac{n!}{s^{n+1}}$	$n \in \mathbb{Z} \geq 0$
t^a	$\frac{\Gamma(a+1)}{s^{a+1}}$	$\mathrm{R}[a] > -1$
e^{at}	$\frac{1}{s-a}$	
$\cos(\omega t)$	$\frac{s}{s^2+\omega^2}$	$\omega \in \mathbb{R}$
$\sin(\omega t)$	$\frac{\omega}{s^2+\omega^2}$	$s >\lvert \mathrm{I}[\omega]\rvert$
$\cosh(\omega t)$	$\frac{s}{s^2-\omega^2}$	$s >\lvert \mathrm{R}[\omega]\rvert$
$\sinh(\omega t)$	$\frac{\omega}{s^2-\omega^2}$	$s >\lvert \mathrm{I}[\omega]\rvert$
$e^{at}\sin(bt)$	$\frac{b}{(s-a)^2+b^2}$	$s > a+\lvert I[b]\rvert$
$e^{at}\cos(bt)$	$\frac{s-a}{(s-a)^2+b^2}$	$b \in \mathbb{R}$
$\delta(t-c)$	e^{-cs}	
$H_c(t)$	$\begin{cases} \frac{1}{s} & \text{for } c \leq 0 \\ \frac{e^{cs}}{s} & \text{for } c > 0 \end{cases}$	
$J_0(t)$	$\frac{1}{\sqrt{s^2+1}}$	
$J_n(at)$	$\frac{\left(\sqrt{s^2+a^2}-s\right)^n}{a^n\sqrt{s^2+a^2}}$	$n \in \mathbb{Z} \geq 0$

In the above table, $J_0(t$ is the zeroth-order Bessel function of the first kind, $\delta(t)$ is the delta function, and $H_c(t)$ is the Heaviside step function.

The Laplace transform has many important properties. The Laplace transform existence theorem states that, if $f(t)$ is piecewise continuous on every finite interval in $(0,\infty)$ satisfying:

$$|f(t)| \le Me^{at}$$

For all $t \in (0,\infty)$, then $\mathcal{L}_t[f(t)](s)$ exists for all $s > a$. The Laplace transform is also unique, in the sense that, given two functions $F_1(t)$ and $F_2(t)$ with the same transform so that:

$$\mathcal{L}_t[F_1(t)](s) = L_t[F_2(t)](s) = f(s),$$

Then Lerch's theorem guarantees that the integral:

$$\int_0^a N(t)dt = 0$$

Vanishes for all $a > 0$ for a null function defined by:

$$N(t) = F_1(t) - F_2(t).$$

The Laplace transform is linear since:

$$\begin{aligned}\mathcal{L}_t[af(t)+bg(t)] &= \int_0^\infty [af(t)+bg(t)]e^{(-st)}dt \\ &= a\int_0^\infty fe^{(-st)}dt - b\int_0^\infty ge^{-st}dt \\ &= a\mathcal{L}_t[f(t)] + b\mathcal{L}_t[g(t)].\end{aligned}$$

The Laplace transform of a convolution is given by:

$$\mathcal{L}_t[f(t) * g(t)] = \mathcal{L}_t[f(t)]L_t[g(t)]$$
$$\mathcal{L}_t^{-1}[FG] = \mathcal{L}_t^{-1}[F] * \mathcal{L}_t^{-1}[G].$$

Now consider differentiation. Let $f(t)$ be continuously differentiable n-1 times in $[0,\infty)$. If $|f(t)| \le Me^{at}$, then:

$$\mathcal{L}_t[f^{(n)}(t)](s) = s^n\mathcal{L}_t[f(t)] - s^{n-1}f(0) - s^{n-2}f'(0) - \ldots - f^{(n-1)}(0).$$

This can be proved by integration by parts,

$$\begin{aligned}\mathcal{L}_t[f'(t)](s) &= \lim_{a\to\infty}\int_0^a e^{-st}f'(t) \\ &= \lim_{a\to\infty}\left\{\left[e^{-st}f(t)\right]_0^a + s\int_0^a e^{-st}f(t)dt\right\} \\ &= \lim_{a\to\infty}\left[e^{-sa}f(a) - f(0) + s\int_0^a e^{-st}f(t)dt\right] \\ &= s\mathcal{L}_t[f(t)] - f(0).\end{aligned}$$

Continuing for higher-order derivatives then gives:

$$\mathcal{L}_t[f''(t)](s) = s^2\mathcal{L}_t[f(t)](s) - sf(0) - f'(0).$$

This property can be used to transform differential equations into algebraic equations, a procedure known as the Heaviside calculus, which can then be inverse transformed to obtain the solution. For example, applying the Laplace transform to the equation:

$$f''(t) + a_1 f'(t) + a_0 f(t) = 0$$

Gives,

$$\begin{aligned}&\left\{s^2\mathcal{L}_t[f(t)](s) - sf(0) - f'(0)\right\} + a_1\left\{s\mathcal{L}_t[f(t)](s) - f(0)\right\}\\&+a_0\mathcal{L}_t[f(t)](s) = 0\\&\mathcal{L}_t[f(t)](s)\left(s^2 + a_1 s + a_0\right) - sf(0) - f'(0) - a_1 f(0) = 0,\end{aligned}$$

which can be rearranged to,

$$\mathcal{L}_t[f(t)](s) = \frac{sf(0) + f'(0) + a_1 f(0)}{s^2 + a_1 s + a_0}$$

If this equation can be inverse Laplace transformed, then the original differential equation is solved.

The Laplace transform satisfied a number of useful properties. Consider exponentiation. If $\mathcal{L}_t[f(t)](s) = F(s)$ for $s > \alpha$ (i.e., $F(s)$ is the Laplace transform of f), then $\mathcal{L}_t[e^{(at)}f](s) = F(s-a)$ for $s > a + \alpha$. This follows from,

$$\begin{aligned}F(s-a) &= \int_0^\infty fe^{-(s-a)t}dt\\&= \int_0^\infty \left[f(t)e^{at}\right]e^{-st}dt\\&= \mathcal{L}_t[e^{at}f(t)](s).\end{aligned}$$

The Laplace transform also has nice properties when applied to integrals of functions. If $f(t)$ is piecewise continuous and $|f(t)| \le M e^{at}$, then,

$$\mathcal{L}_t\left[\int_0^t f(t')dt'\right] = \frac{1}{s}\mathcal{L}_t[f(t)](s).$$

Laplace Transform Properties

We are aware that the Laplace transform of a continuous signal $x(t)$ is given by,

$$x(s) = \int_{-\infty}^{\infty} x(t)e^{-st}dt$$

And inverse Laplace transform is given by:

$$x(t) = \frac{1}{2\pi j} \int_{\sigma - j\infty}^{\sigma + j\infty} X(s) e^{st} ds$$

The Properties of Laplace transform simplifies the work of finding the s-domain equivalent of a time domain function when different operations are performed on signal like time shifting, time scaling, time reversal etc. These properties also signify the change in ROC because of these operations.

These properties are also used in applying Laplace transform to the analysis and characterization of LTI systems.

Linearity of the Laplace Transform

Statement:

If $x_1(t) \overset{\mathcal{L}}{\leftrightarrow} X_1(s)$ with a region of convergence denoted as R_1

and $x_2(t) \overset{\mathcal{L}}{\leftrightarrow} X_2(s)$ with a region of convergence denoted as R_2

then $ax_1(t) + bx_2(t) \overset{\mathcal{L}}{\leftrightarrow} aX_1(s) + bX_2(s)$, with ROC containing $R_1 \cap R_2$

Proof:

Consider the linear combination of two signals $x_1(t)$ and $x_2(t)$ as $z(t) = ax_1(t) + bx_2(t)$. Now, take the Laplace transform of $z(t)$ as:

$$\mathcal{L}\{z(t)\} = \mathcal{L}\{ax_1(t) + bx_2(t)\} = \int_{-\infty}^{\infty} \{ax_1(t) + bx_2(t)\} e^{-st} dt$$

$$= a\int_{-\infty}^{\infty} x_1(t) e^{-st} dt + b\int_{-\infty}^{\infty} x_2(t) e^{-st} dt$$

$$= aX_1(s) + bX_2(s)$$

The resulting ROC is as large as the region in common between the independent ROCs. However, there may be pole-zero cancellation in the linear combination, which results in extending the ROC beyond the common region.

In this example, we illustrate the fact that the ROC for the Laplace transform of a linear combination of signals can sometimes extend beyond the intersection of the ROCs for the individual terms. Consider,

$$x(t) = x_1(t) - x_2(t)$$

where the Laplace transforms of $x_1(t)$ and $x_2(t)$ are, respectively:

$$X_1(s) = \frac{1}{s+1}, \ \mathrm{Re}\{s\} > -1 \quad \text{and} \quad X_2(s) = \frac{1}{(s+1)(s+2)}, \ \mathrm{Re}\{s\} > -1$$

The pole-zero plot, including the ROCs for $X_1(s)$ and $X_1(s)$ v, is shown below respectively in figure (a) and (b).

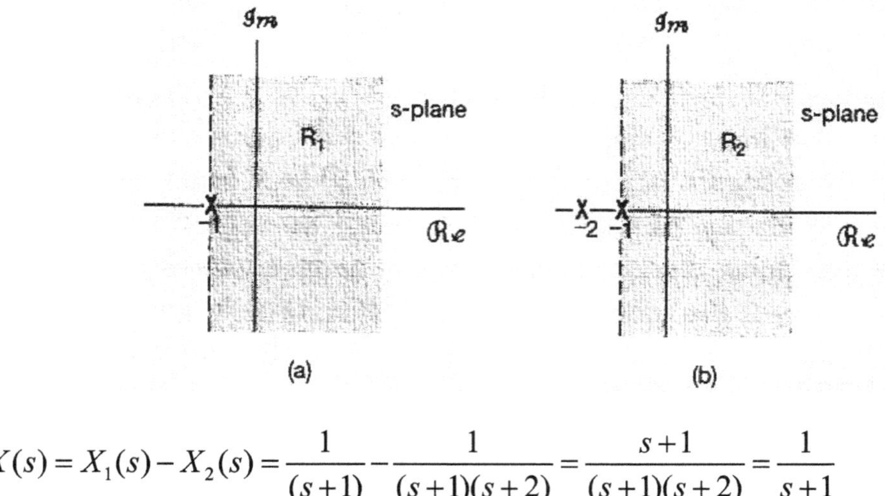

$$X(s) = X_1(s) - X_2(s) = \frac{1}{(s+1)} - \frac{1}{(s+1)(s+2)} = \frac{s+1}{(s+1)(s+2)} = \frac{1}{s+1}$$

Thus, in the linear combination of $x_1(t)$ and $x_1(t)$, the pole t s=-1 is cancelled by a zero at $s=-1$. The pole-zero plot for $X(s)$ is shown below.

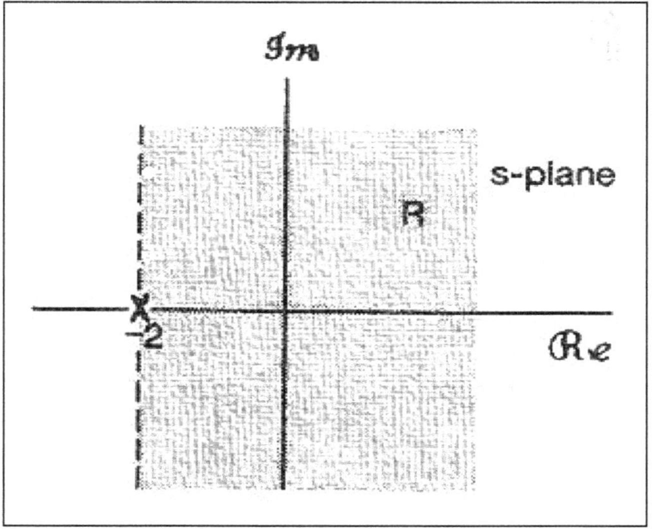

The intersection of ROCs for $X_1(s)$ and $X_2(s)$ is $\text{Re}\{s\} > -1$. However, since the ROC is always bounded by a pole or infinity, for this example the ROC for $X(s)$ can be extended to the left to be bounded by the pole at $s=-2$, because of the pole-zero cancellation at $s=-1$.

Time Shifting

Statement:

If $x(t) \overset{\mathcal{L}}{\leftrightarrow} X(s)$ with ROC=R

then $x(t-\tau) \overset{\mathcal{L}}{\leftrightarrow} e^{-st} X(s)$ with ROC=R

Proof:

$$\mathcal{L}\{x(t-\tau)\} = \int_{-\infty}^{\infty} x(t-\tau)e^{-st}\,dt$$

$$\text{Let } t-\tau = p$$

$$= \int_{-\infty}^{\infty} x(p)e^{-s(p+\tau)}\,dt$$

$$= e^{-st}\int_{-\infty}^{\infty} x(p)e^{-sp}\,dt$$

$$= e^{-st}X(s)$$

Illustration:

As productof $X(s)$ with e^{-st} will not effect the poles of $X(s)$, ROC remains unaltered.

Shifting in S-domain

Statement:

If $x(t)\overset{\mathcal{L}}{\leftrightarrow}X(s)$ with $ROC = R$

then $e^{-s_0 t}x(t)\overset{\mathcal{L}}{\leftrightarrow}X(s-s_0)$ with $ROC = R + \mathrm{Re}\{s_o\}$

Proof:

$$\mathcal{L}\{e^{s_0 t}x(t)\} = \int_{-\infty}^{\infty} e^{s_0 t}x(t)e^{-st}\,dt$$

$$= \int_{-\infty}^{\infty} x(t)e^{-(s-s_0)t}\,dt$$

$$= X(s-s_0)$$

That is, the ROC associated with $X(s-s_0)$ is that of $X(s)$, shifted by $\mathrm{Re}\{s_o\}$.Thus, for any value s that is in R, the value $s+\mathrm{Re}\{s_o\}$ will be in ℜ1.This is illustrated in figure below. Figure (a) and (b) represents ROC of $X(s)$ and $X(s-s_0)$ respectively.

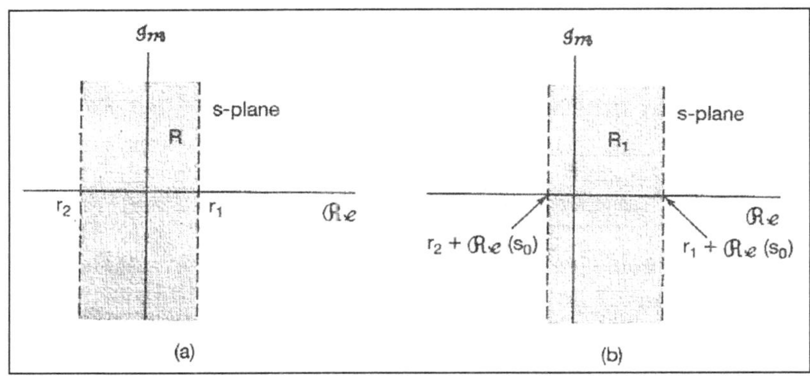

Note that if $X(s)$ has a pole or zero at s=a, then $X(s-s_0)$ has a pole or zero at s-so=a, i.e., $s-s_0=a$.

A special case is observed when $s_0 = j\omega_o$, i.e., when a signal is used to modulate a periodic complex exponential $e^{j\omega_o t}$.

In this case $e^{j\omega_o t}x(t) \overset{\mathcal{L}}{\leftrightarrow} X(s-j\omega_o)$ with ROC = R.

This is true because ROC depends on real part of 's' not the imaginary part.

Time Scaling

Statement:

If $x(t) \overset{\mathcal{L}}{\leftrightarrow} X(s)$ with ROC = R m

then $x(at) \overset{\mathcal{L}}{\leftrightarrow} \frac{1}{|a|}X\left(\frac{s}{a}\right)$ with ROC = $R_1 = aR$

Proof:

To prove this we have to consider two cases: a (real) is positive and a is negative. Case: For $a>0$:

$$\mathcal{L}\{x(at)\} = \int_{-\infty}^{\infty} x(at)e^{-st}dt$$

Using the substitution of $\lambda = at$; $dt = ad\lambda$.

$$= \frac{1}{a}\int_{-\infty}^{\infty} x(\lambda)e^{-\left(\frac{s}{a}\right)\lambda}d\lambda$$

$$= \frac{1}{a}X\left(\frac{s}{a}\right)$$

Case: For $a<0$.

$$\mathcal{L}\{x(at)\} = \int_{-\infty}^{\infty} x(at)e^{-st}dt$$

Using the substitution of $\lambda = at$; $dt = ad\lambda$.

$$= \frac{1}{a}\int_{-\infty}^{\infty} x(\lambda)e^{-\left(\frac{s}{a}\right)\lambda}d\lambda$$

$$= -\frac{1}{a}X\left(\frac{s}{a}\right)$$

Combining the two cases, we get $x(at) \overset{\mathcal{L}}{\leftrightarrow} \frac{1}{|a|}X\left(\frac{s}{a}\right)$ with ROC = $R_1 = aR$.

Illustration: For example Laplace transform of $x(t) = e^{bt}u(t)$ is $X(s) = \frac{1}{s+b}$ with

ROC: $\text{Re}\{s\} = \sigma > -b$ representing right-sided signal. Then $\frac{1}{|a|}X\left(\frac{s}{a}\right) = \frac{1}{|a|}\frac{1}{\left(\frac{s}{a}+b\right)}$ with

ROC: $\text{Re}\{s\} > a(-b) \Rightarrow \text{Re}\{s\} > a(-b)$ representing ROC: $R_1 = aR$.

Let the ROC $X(s)$ is given as shown below.

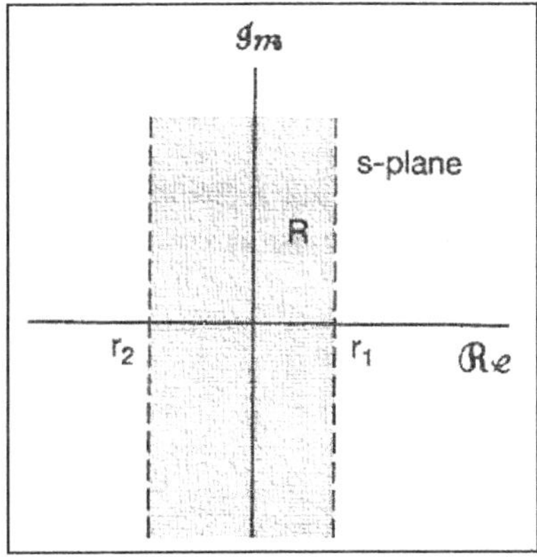

Change in ROC is also explained with different ranges of a.

- If $a > 1$ then the resultant ROC is expanded.

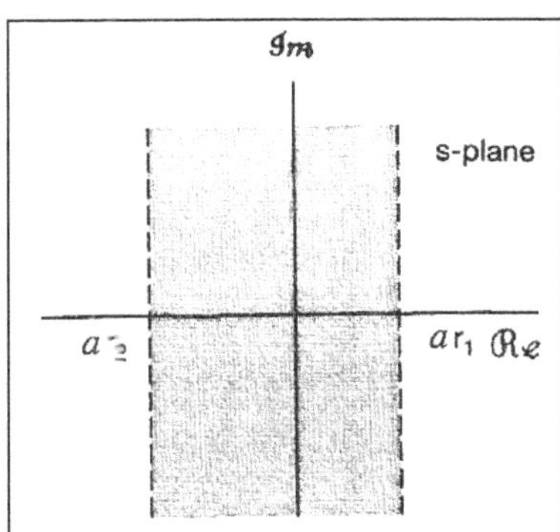

- If $a < -1$ then the resultant ROC expands and the bounds get reversed.

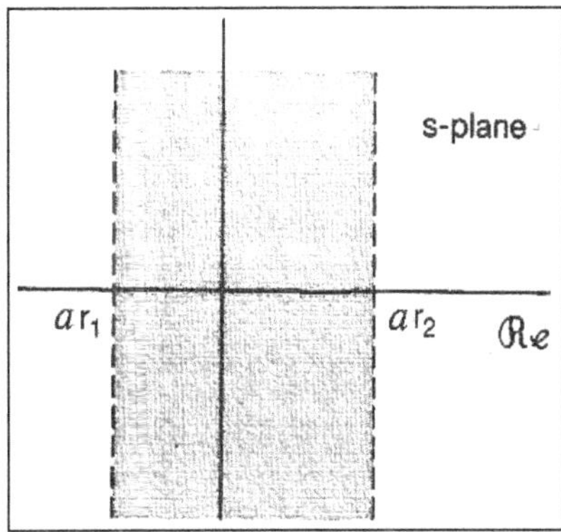

- If $0 < a < 1$ then the resultant ROC is compressed.

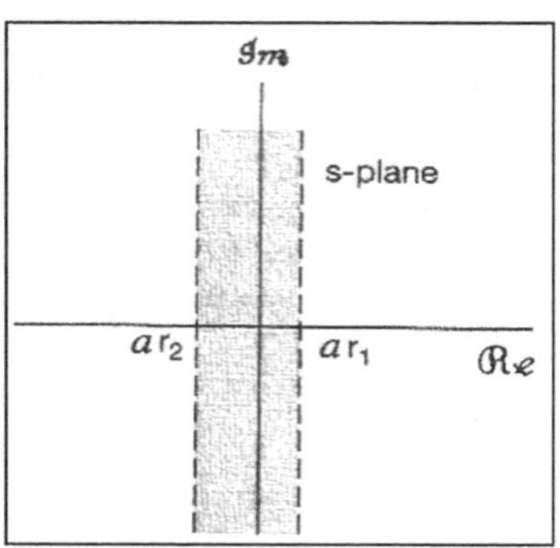

- If $-1 < a < 1$ then the resultant ROC is compressed and the bounds get reversed.

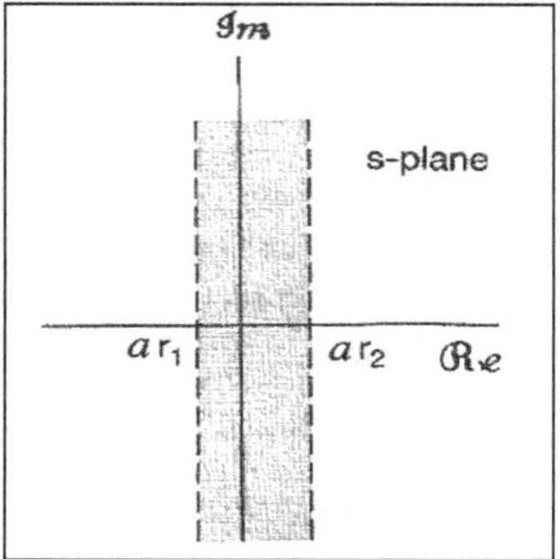

- If $a = -1$, then it gives rise to Time Reversal operation with the statement $x(-t) \overset{\mathcal{L}}{\leftrightarrow} X(-s)$ with $\mathrm{R}_1 = -\mathrm{R}$.

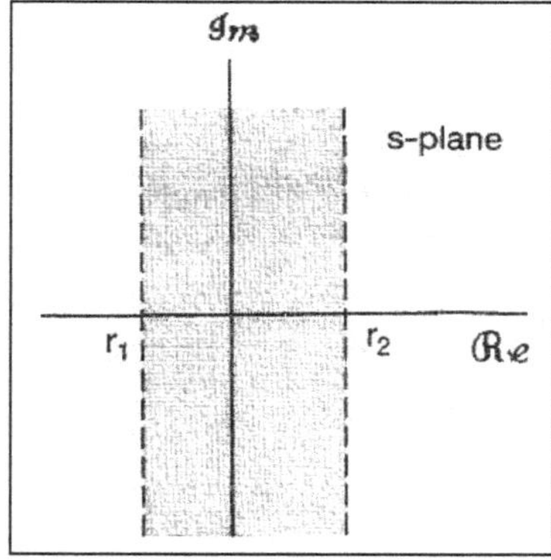

Conjugation

Statement:

If $x(t) \overset{\mathcal{L}}{\leftrightarrow} X(s)$ with $\mathrm{ROC} = \mathrm{R}$

Then $x^*(t) \overset{\mathcal{L}}{\leftrightarrow} X^*(s^*)$ with $\mathrm{ROC} = \mathrm{R}$

$$\mathcal{L}\{x^*(t)\} = \int_{-\infty}^{\infty} x^*(t) e^{-st} dt$$

As we know that $s = \sigma + j\omega$,

$$= \int_{-\infty}^{\infty} x^*(t) e^{-\sigma t} e^{-j\omega t} dt$$

$$= \left(\int_{-\infty}^{\infty} x(t) e^{-\sigma t} e^{-j\omega t} dt \right)$$

$$= \left(\int_{-\infty}^{\infty} x(t) e^{-(\sigma - j\omega)t} dt \right)^*$$

$$\left(\int_{-\infty}^{\infty} x(t) e^{-(s^*)t} dt \right)^*$$

$$= \left(X(s^*) \right)^* = X^*(s^*)$$

Also $X(s) = X^*(s^*)$ when $x(t)$ is real.

Illustration: If $x(t)$ is real then and if $X(s)$ has a pole or zero at $\mathrm{s} = \mathrm{so}$, then $X(s)$ also has a pole or zero at the complex conjugate point $\mathrm{s} = \mathrm{s}_o^*$. As only imaginary part changes and not the real part, ROC remains unaltered.

Convolution Property

Statement:

If $x_1(t) \overset{\mathcal{L}}{\leftrightarrow} X_1(s)$ with ROC=R_1

and $x_2(t) \overset{\mathcal{L}}{\leftrightarrow} X_2(s)$ with ROC=R_2

Then $x_1(t) * x_2(t) \overset{\mathcal{L}}{\leftrightarrow} X_1(s).X_2(s)$, with ROC containing $R_1 \cap R_2$.

Proof:

$$\mathcal{L}\{z(t)\} = \mathcal{L}\{x_1(t) * x_2(t) = \int_{-\infty}^{\infty} \{x_1(t) * x_2(t)\} e^{-st} dt$$

$$= \int_{-\infty}^{\infty} \left\{ \int_{-\infty}^{\infty} x_1(\tau) x_2(t - \tau) d\tau \right\} e^{-st} dt$$

Interchanging the order of integrations:

$$\mathcal{L}\{x_1(t) * x_2(t)\} = \int_{-\infty}^{\infty} x_1(\tau) \left\{ \int_{-\infty}^{\infty} x_2(t - \tau) e^{-st} dt \right\} d\tau$$

$$= \int_{-\infty}^{\infty} x_1(\tau) \{ e^{-s\tau} X_2(s) \} d\tau \text{(Since from time shifting property)}$$

$$=X_2(s)\int_{-\infty}^{\infty} x_1(\tau)e^{-st}d\tau$$

$$= X_1(s).X_2(s)$$

In a manner, like the linearity property, the ROC of $X_1(s).X_2(s)$ includes the intersection of the ROCs of $X_1(s)$ and $X_2(s)$ and may be larger if pole-zero cancellation occurs in the product.

For example: If,

$$X_1(s) = \frac{s+1}{s+2}, \text{Re}\{s\} > -2 \text{ and } X_2(s) = \frac{s+2}{s+1}, \text{Re}\{s\} > -1,$$

then $X_1(s).X_2(s) = 1$, and its ROC is the entire s-plane.

This property plays an important role in the analysis of LTI systems.

Region of Convergence

The region of convergence, known as the ROC, is important to understand because it defines the region where the Laplace transform exists. The Laplace transform of a sequence is defined as:

$$H(s) = \int_{-\infty}^{\infty} h(t)e^{-(st)}dt$$

The ROC for a given $h(t)$, is defined as the range of t for which the Laplace transform converges. If we consider a causal, complex exponential, $h(t) = e^{-(at)}u(t)$, we get the equation,

$$\int_0^{\infty} e^{-(at)}e^{-(st)}dt = \int_0^{\infty} e^{-((a+s)t)}dt$$

Evaluating this, we get:

$$\frac{-1}{s+a}\left(\underset{t\to\infty}{\text{limit}}\, e^{-((s+a)t)} - 1\right)$$

Notice that this equation will tend to infinity when $\underset{t\to\infty}{\text{limit}}\, e^{-((s+a)t)}$ tends to infinity. To understand when this happens, we take one more step by using $s = \sigma + i\omega$ to realize this equation as:

$$\underset{t\to\infty}{\text{limit}}\, e^{-(i\omega t)}e^{-((\sigma+a)t)}$$

Recognizing that $e^{-(i\omega t)}$ is sinusoidal, it becomes apparent that $e^{-((\sigma+a)t)}$ is going to determine whether this blows up or not. What we find is that if $\sigma + a$ is positive, the exponential will be to a negative power, which will cause it to go to zero as t tends to infinity. On the other hand, if $\sigma + a$ is negative or zero, the exponential will not be to a negative power, which will prevent it from tending to zero and the system will not converge. What all of this tells us is that for a causal signal, we have convergence when.

Condition for Convergence

$$\Re(s) > -a$$

Alternatively, we can note that since the Laplace transform is a power series, it converges when $h(t)e^{-(st)}$ is absolutely summable. Therefore,

$$\int_{-\infty}^{\infty} h(t)e^{-(st)}dt < \infty$$

must be satisfied for convergence.

Although we will not go through the process again for anticausal signals, we could. In doing so, we would find that the necessary condition for convergence is when.

Necessary Condition for Anti-causal Convergence

$$\Re(s) < -a$$

Properties of the Region of Convergence

The Region of Convergence has a number of properties that are dependent on the characteristics of the signal, $h(t)$.

The ROC cannot contain any poles. By definition a pole is a where $H(s)$ is infinite. Since $H(s)$ must be finite for all s for convergence, there cannot be a pole in the ROC.

If $h(t)$ is a finite-duration sequence, then the ROC is the entire s-plane, except possibly $s = 0$ or $|s| = \infty$. A finite-duration sequence is a sequence that is nonzero in a finite interval $t_1 \leq t \leq t_2$. As long as each value of $h(t)$ is finite then the sequence will be absolutely summable. When $t_2 > 0$ there will be a s^{-1} term and thus the ROC will not include $s = 0$. When $t_1 < 0$ then the sum will be infinite and thus the ROC will not include |s|=∞. On the other hand, when $t_2 \leq 0$ then the ROC will include $s = 0$, and when $t_1 \geq 0$ the ROC will include $|s| = \infty$. With these constraints, the only signal, then, whose ROC is the entire z-plane is $h(t) = c\delta(t)$.

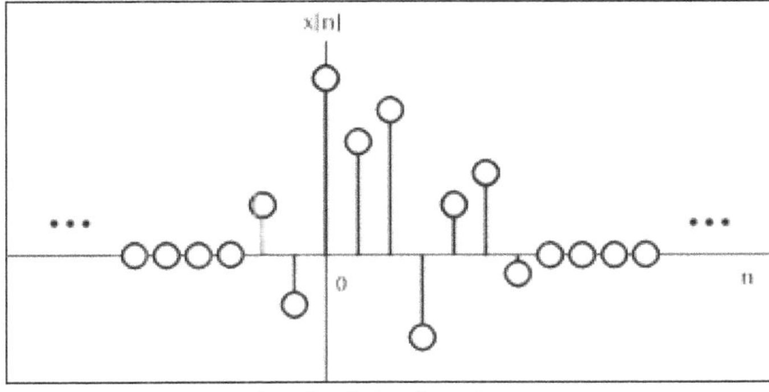

An example of a finite duration sequence.

The next properties apply to infinite duration sequences. As noted above, the z-transform converges when $|H(s)|<\infty$. So we can write,

$$|H(s)|=\left|\int_{-\infty}^{\infty} h(t)e^{-(st)}dt\right| \leq \int_{-\infty}^{\infty} \left|h(t)e^{-(st)}\right| dt = \int_{-\infty}^{\infty} |h(t)|\left|e^{-(st)}\right| dt$$

We can then split the infinite sum into positive-time and negative-time portions. So,

$$|H(s)| \leq N(s)+P(s)$$

where,

$$N(s)=\int_{-\infty}^{-1} |h(t)|\left|e^{-(st)}\right| dt$$

and

$$P(s)=\int_{0}^{\infty} |h(t)|\left|e^{-(st)}\right| dt$$

In order for $|H(s)|$ to be finite, $|h(t)|$ must be bounded. Let us then set,

$$|h(t)| \leq C_1 r_1^t$$

for,

$$t<0$$

$$|h(t)| \leq C_2 r_2^t$$

for,

$$t \geq 0$$

From this some further properties can be derived:

If $h(t)$ is a right-sided sequence, then the ROC extends outward from the outermost pole in $H(s)$. A right-sided sequence is a sequence where $h(t)=0$ for $t<t_1<\infty$. Looking at the positive-time portion from the above derivation, it follows that,

$$P(s)=C_2\int_{0}^{\infty} r_2^t \left|e^{-(st)}\right| dt = C_2\int_{0}^{\infty} \frac{r_2}{\left|e^{-(st)}\right|} dt$$

Thus in order for this integral to converge, $|e^{-s}|>r_2$, and therefore the ROC of a right-sided sequence is of the form $|e^{-s}|>r_2$.

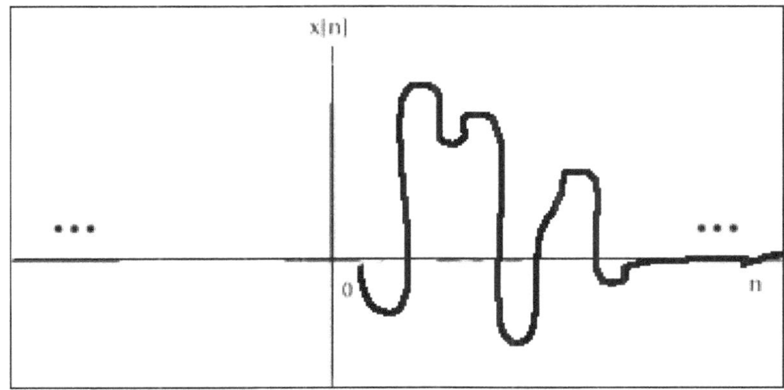

A right-sided sequence.

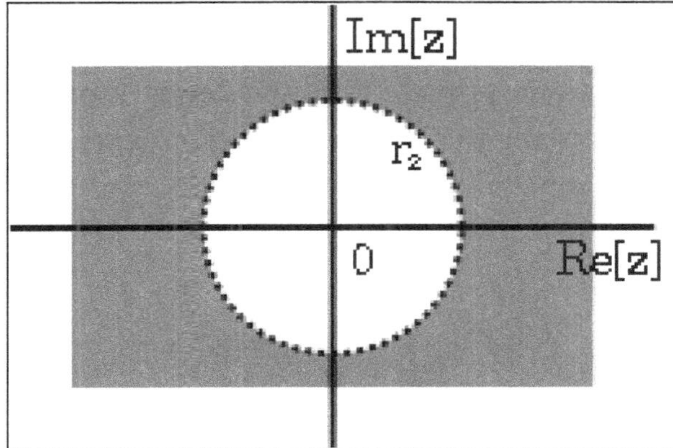

The ROC of a right-sided sequence.

If $h(t)$ is a left-sided sequence, then the ROC extends inward from the innermost pole in $H(s)$. A left-sided sequence is a sequence where $h(t) = 0$ for $t > t_1 > -\infty$. Looking at the negative-time portion from the above derivation, it follows that,

$$P(s) \leq C_1 \int_{-\infty}^{-1} r_1^t e^{-(st)} dt = C_1 \int_{-\infty}^{-1} \left(\frac{r_1}{|e^{-s}|} \right)^t dt = C_1 \int_1^{\infty} \left(\frac{|e^{-s}|}{r_1} \right)^k dk$$

Thus in order for this integral to converge, $|e^{-s}| < r_1$, and therefore the ROC of a left-sided sequence is of the form $|e^{-s}| < r_1$.

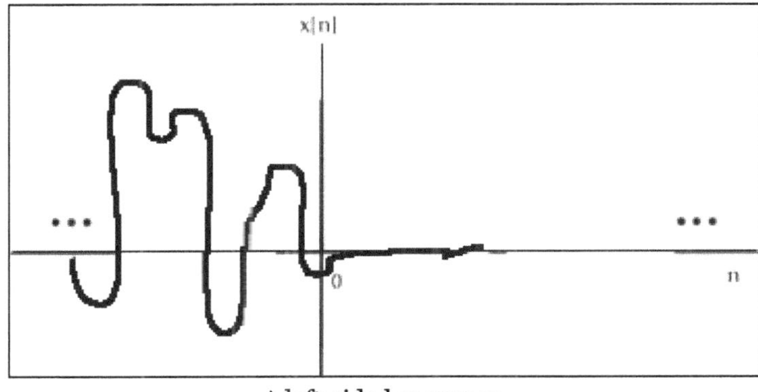

A left-sided sequence.

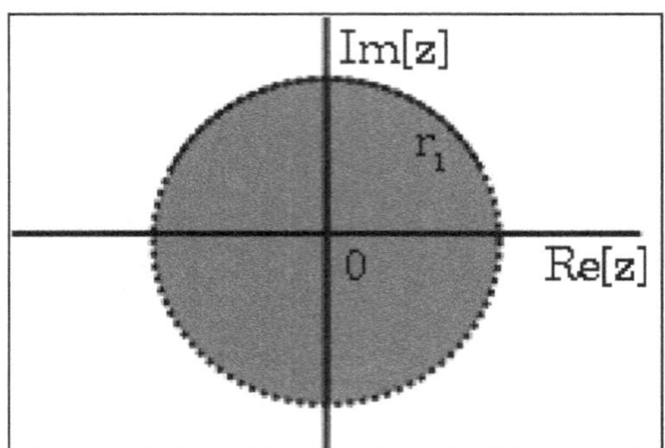

The ROC of a left-sided sequence.

If $h(t)$ is a two-sided sequence, the ROC will be a ring in the z-plane that is bounded on the interior and exterior by a pole. A two-sided sequence is an sequence with infinite duration in the positive and negative directions. From the derivation of the above two properties, it follows that if $r_2 < |e^{-s}| < r_2$ converges, then both the positive-time and negative-time portions converge and thus $H(s)$ converges as well. Therefore the ROC of a two-sided sequence is of the form $-r_2 < |e^{-s}| < r_2$.

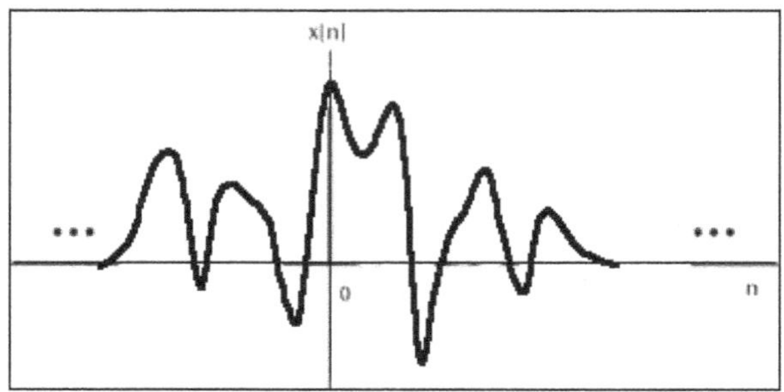

A two-sided sequence.

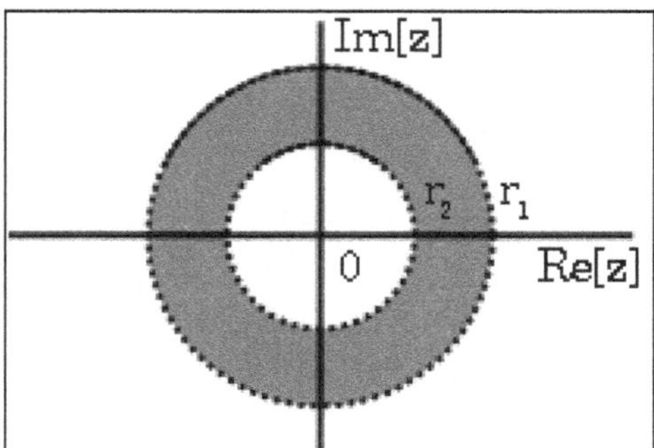

The ROC of a two-sided sequence.

Examples:

To gain further insight it is good to look at a couple of examples.

Example:

Lets take:

$$h_1(t) = \left(\frac{1}{2}\right)^t u(t) + \left(\frac{1}{4}\right)^t u(t)$$

The Laplace-transform of $\left(\frac{1}{2}\right)^t u(t)$ is $\frac{s}{s-\frac{1}{2}}$ with an ROC at $|s| > \frac{1}{2}$.

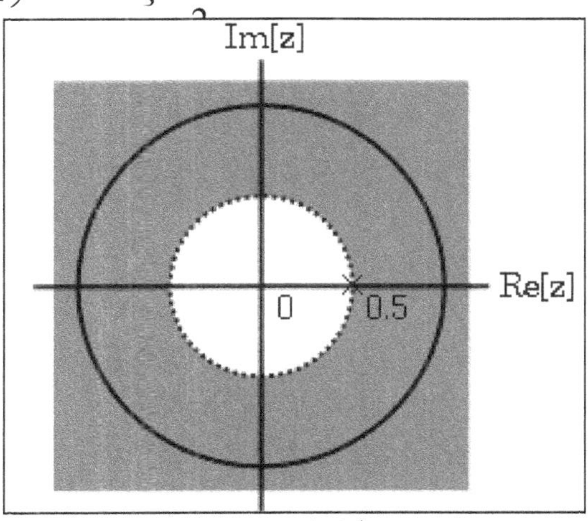

The ROC of $\left(\frac{1}{2}\right)^t u(t)$

The z-transform of $\left(\frac{-1}{4}\right)^t u(t)$ is $\frac{s}{s+\frac{1}{4}}$ with an ROC at $|s| > \frac{-1}{4}$.

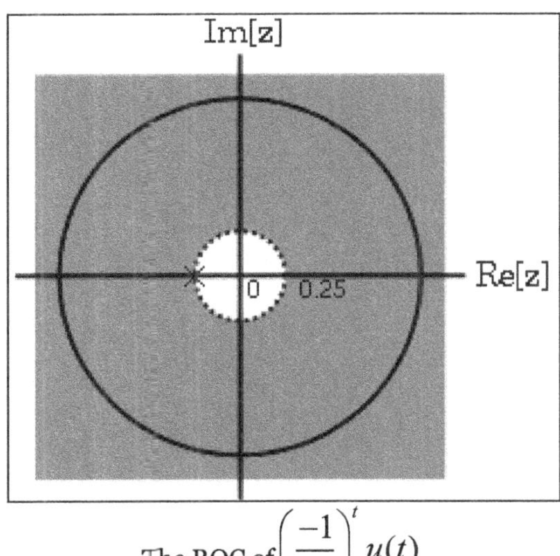

The ROC of $\left(\frac{-1}{4}\right)^t u(t)$

Due to linearity,

$$H_1(s) = \frac{s}{s-\frac{1}{2}} + \frac{s}{s+\frac{1}{4}}$$

$$= \frac{2s\left(s-\frac{1}{8}\right)}{\left(s-\frac{1}{2}\right)\left(s+\frac{1}{4}\right)}$$

By observation it is clear that there are two zeros, at 0 and $\frac{1}{8}$, and two poles, at $\frac{1}{2}$, and $\frac{-1}{4}$. Following the above properties, the ROC is $|s| > \frac{1}{2}$.

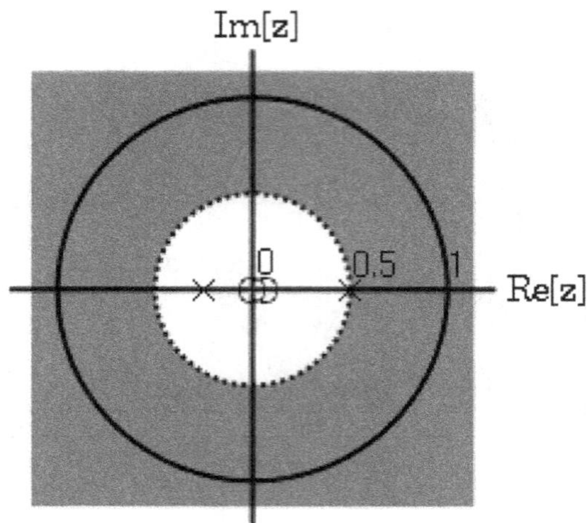

The ROC of $h_1(t) = \left(\frac{1}{2}\right)^t u(t) + \left(\frac{-1}{4}\right)^t u(t)$

Example:

Now take:

$$h_2(t) = \left(\frac{-1}{4}\right)^t u(t) - \left(\frac{1}{2}\right)^t u((-t)-1)$$

The z-transform and ROC of $\left(\frac{-1}{4}\right)^t u(t)$ was shown in the example above. The Laplace-transorm of $\left(-\left(\frac{1}{2}\right)^t\right)u((-t)-1)$ is $\frac{s}{s-\frac{1}{2}}$ with an ROC at $|s| > \frac{1}{2}$.

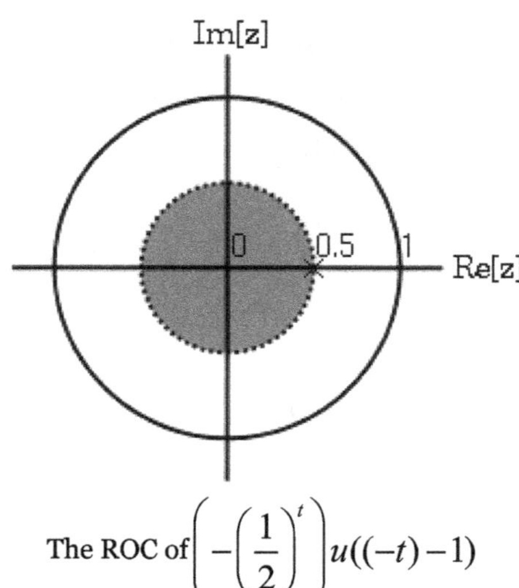

The ROC of $\left(-\left(\frac{1}{2}\right)^t\right)u((-t)-1)$

Once again, by linearity,

$$H_2(s) = \frac{s}{s+\frac{1}{4}} + \frac{s}{s-\frac{1}{2}}$$
$$= \frac{s\left(2s-\frac{1}{8}\right)}{\left(s+\frac{1}{4}\right)\left(s-\frac{1}{2}\right)}$$

By observation it is again clear that there are two zeros, at 0 and $\frac{1}{16}$, and two poles, at $\frac{1}{2}$, and $\frac{-1}{4}$. in this case though, the ROC is $|s| < \frac{1}{2}$.

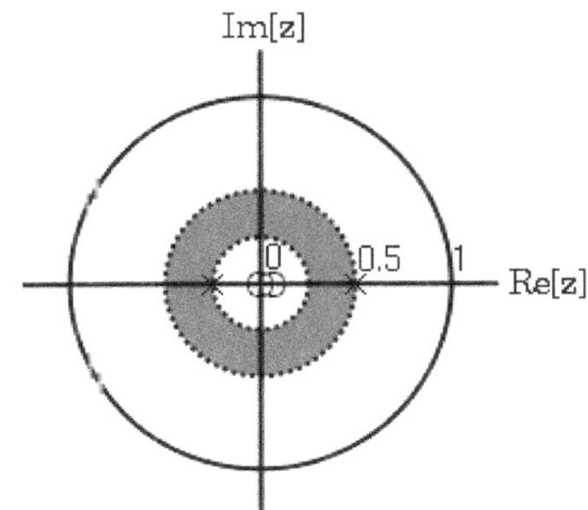

The ROC of $h_2(t) = \left(\frac{-1}{4}\right)^t u(t) - \left(\frac{1}{2}\right)^t u((-t)-1)$.

The Laplace Transform of a Function

The Laplace Transform of a function $y(t)$ is defined by:

$$Y(s) = L[y(t)](s) = \int_0^\infty e^{-(st)} y(t)dt$$

if the integral exists. The notation $L[y(t)](s)$ means take the Laplace transform of $y(t)$. The functions $y(t)$ and $Y(s)$ are partner functions. Note that $Y(s)$ is indeed only a function of s since the definite integral is with respect to t.

Examples:

Let $y(t) = exp(t)$. We have:

$$Y(s) = \int_0^\infty e^{-(st)} e^t dt = \int_0^\infty e^{-(s-1)t} dt \frac{1}{s-1}$$

The integral converges if $s>1$. The functions $exp(t)$ and $1/(s-1)$ are partner functions.

Let $y(t)=\cos(3t)$. We have,

$$Y(s)=\int_0^{\infty} e^{-(st)}\cos(3t)dt=\frac{1}{s^2+9}$$

The integral converges for $s>0$. The integral can be computed by doing integration by parts twice or by looking in an integration table.

Existence of the Laplace Transform

If $y(t)$ is piecewise continuous for $t>=0$ and of exponential order, then the Laplace Transform exists for some values of s. A function $y(t)$ is of exponential order c if there is exist constants M and T such that,

$$|y(t)|\le Me^{ct} \quad t\ge T$$

All polynomials, simple exponentials ($\exp(at)$, where a is a constant), sine and cosine functions, and products of these functions are of exponential order. An example of a function not of exponential order is $\exp(t\hat{\ }2)$. This function grows too rapidly. The integral:

$$\int_0^{\infty} e^{-st}e^{t^2}dt$$

does not converge for any value of s.

Table of Laplace Transforms

The following table lists the Laplace Transforms for a selection of functions.

Function $y(t)$	Transform $Y(s)$	s
1	1/s	s>0
t	1/s^2	s>0
t^n, n=integer	n!/s^(n+1)	s>0
exp(at), a=constant	1/(s-a)	s>a
cos(bt), b=constant	s/(s^2+b^2)	s>0
sin(bt), b=constant	b/(s^2+b^2)	s>0
exp(at) cos(bt)	(s-a)/[(s-a)^2+b^2]	s>a
exp(at) sin(bt)	b/[(s-a)^2+b^2]	s>a

Rules for Computing Laplace Transforms of Functions

There are several formulas and properties of the Laplace transform which can greatly simplify calculation of the Laplace transform of functions.

Linearity

Like differentiation and integration the Laplace transformation is a linear\ operation. What does this mean? In words, it means that the Laplace transform of a constant times a function is the constant times the Laplace transform of the function. In addition the Laplace transform of a sum of functions is the sum of the Laplace transforms.

Let us restate the above in mathspeak. Let c and Y_1(s) denote the Laplace transforms of y_1(t) and y_2(t), respectively, and let c_1 be a constant. Recall that L[f(t)](s) denotes the Laplace transform of f(t). We have,

$$L[c_1 y_1(t)](s) = c_1 L[y_1(t)](s) = c_1 Y_1(s)$$

$$L[y_1(t) + y_2(t)](s) = L[y_1(t)](s) + L[y_2(t)](s) = Y_1(s) + Y_2(s)$$

As a corollary, we have the third formula:

$$L[c_1 y_1(t) + c_2 y_2(t)](s) = c_1 Y_1(s) + c_2 Y_2(s)$$

Here are several examples:

$$L[3e^t](s) = 3L[e^t](s) = 3 \cdot \frac{1}{s-1} = \frac{3}{s-1}$$

Here we have used the results in the table for the Laplace transform of the exponential. Here are a couple of more examples:

$$L[e^t + \cos 2t](s) = L[e^t](s) + L[\cos 2t](s) = \frac{1}{s-1} + \frac{s}{s^2+4}$$

$$L[3e^t + 4\cos 2t](s) = 3L[e^t](s) + 2L[\cos 2t](s) = \frac{3}{s-1} + \frac{4s}{s^2+4}$$

Translation Property

The translation formula states that $Y(s)$ is the Laplace transform of $y(t)$, then,

$$L[e^{2t}t](s) = Y(s-2) = \frac{1}{(s-2)^2}$$

Laplace Transform of the Derivative

Suppose that the Laplace transform of $y(t)$ is $Y(s)$. Then the Laplace Transform of $y'(t)$ is,

$$L[y'(t)](s) = sY(s) - y(0)$$

For the second derivative we have,

$$L[y''(t)](s) = s^2Y(s) - sy(0) - y'(0)$$

For the n'th derivative we have,

$$L[y^{(n)}(t)](s) = s^nY(s) - s^{n-1}y(0) - s^{n-2}y'(0) - \ldots - y^{(n-1)}(0)$$

Derivatives of the Laplace Transform

Let Y(s) be the Laplace Transform of $y(t)$. Then,

$$L[t^n y(t)](s) = (-1)^n \frac{d^nY}{ds^n}(s)$$

Here is an example. Suppose we wish to compute the Laplace transform of $t\sin(t)$. The Laplace transform of $\sin(t)$ is $1/(s\hat{}2+1)$. Hence, we have,

$$L[t\sin t] = -\frac{d}{ds}[\frac{1}{s^2+1}] = \frac{2s}{(s^2+1)^2}$$

Existence of Laplace Transform

The Laplace transform $L[f(x)]$ exists provided the integral,

$$\int_0^\infty f(x)e^{-px}dx = \lim_{a\to\infty}\int_0^\infty f(x)e^{-px}dx$$

exists for sufficiently large p.

Preliminary

Absolute Convergence

If the integral,

$$\int_a^b |f(x)|dx$$

converges, then the integral,

$$\int_a^b f(x)dx$$

converges absolutely. Note that it is okay for a, b to be $\pm\infty$.

Comparison Test

If $|f(x)| \leq g(x)$ for all $a \leq x \leq b$ and the integral,

$$\int_a^b g(x)dx$$

converges, then the integral,

$$\int_a^b f(x)dx$$

also converges absolutely.

Triangle Inequality

$$\left|\int_a^b f(x)dx\right| \leq \int_a^b |f(x)|dx$$

Exponential Order

The function $f(x)$ is said to have exponential order if there exist constants M, c, and n such that,

$$|f(x)| \leq Me^{cx}$$

for all $x \geq n$.

Criteria for Convergence (I)

The Laplace transform $L[f(x)]$ exists if it has exponential order,

$$\int_0^b |f(x)|dx$$

exists for any $b > 0$. Since we only need to show convergence for sufficiently large p, assume $p > c$ and $p > 0$.

$$\begin{aligned}\int_0^\infty |f(x)e^{-px}|dx &= \int_0^n |f(x)e^{-px}|dx + \int_n^\infty |f(x)e^{-px}|dx \\ &\leq \int_0^n |f(x)|dx + \int_n^\infty e^{-px}|f(x)|dx \qquad 0 < e^{-px} \leq 1 \\ &\leq \int_0^n |f(x)|dx + \int_n^\infty e^{-px}Me^{cx}dx \qquad \text{exponential order}\end{aligned}$$

$$=\int_0^n |f(x)|\,dx + M\left[\frac{e^{(c-p)x}}{c-p}\right]_n^{\infty} \qquad p > c$$

$$=\int_0^n |f(x)|\,dx + M\frac{e^{(c-p)n}}{p-c}$$

The first integral exists by assumption, and the second term is finite for $p > c$, so the integral,

$$\int_0^{\infty} f(x)e^{-px}dx$$

converges absolutely and the Laplace transform $L[f(x)]$ exists.

Criteria for Convergence (II)

The Laplace transform $L[f(x)]$ exists if:

- $f(x)$ has exponential order,
- On every closed interval $[0,\ b]$,
- $f(x)$ is bounded,
- $f(x)$ is piecewise continuous,
- $f(x)$ has at most a finite number of discontinuities.

Requirements $2(a-c)$ imply that,

$$\int_0^b |f(x)|\,dx$$

will always exist, so we automatically satisfy criterion (I).

$$F(p) \rightarrow 0 \text{ as } p \rightarrow \infty$$

Assume $f(x)$ satisfies criterion (I) This implies $F(p) = L[f(x)]$ will exist if if $p \geq m$ for some m. I want to show that $|F(p)|$ can be made arbitrarily close to 0 for sufficiently large p. Choose an $\in > 0$. Fix a p. We will discover how large p needs to be as we go; we only care about $p \rightarrow \infty$, so we may choose p to be as large as we need.

$$|F(p| = \left|\int_0^{\infty} f(x)e^{-px}dx\right| \leq \int_0^{\infty} \left|f(x)e^{-px}\right| dx = G(p).$$

Note that as $p \rightarrow \infty$, $e^{-px} \rightarrow 0$ for $x > 0$, so that I should be able to make the integral arbitrarily small for large p. The only potential complication is near $x = 0$, so we will need to deal with that separately. The important point here is that the part near 0 does not contribute very much to the integral. Let,

$$K_a(p) = \int_a^{\infty} \left|f(x)e^{-px}\right| dx$$

Then, $G(p) = \lim_{a \to 0^+} K_a(p)$. By the definition of a limit, there exists an $\delta > 0$ such that,

$$|K_a(p) - F(p)| < \frac{\in}{2} \quad \text{for all } 0 < a \le \delta.$$

Using this (with a = δ),

$$\int_0^\delta |f(x)e^{-px}| dx = F(p) - K_\delta(P) < \frac{\in}{2}$$

If I assume $p > m$, then,

$$\begin{aligned}
|F(p)| &< \frac{\in}{2} + \int_\delta^\infty |f(x)e^{-px}| dx \\
&= \frac{\in}{2} + \int_\delta^\infty |f(x)| e^{-(p-n)x} e^{-nx} dx \\
&\le \frac{\in}{2} + \int_\delta^\infty |f(x)| e^{-(p-n)\delta} e^{-nx} dx \quad \text{since } x \ge \delta \\
&= \frac{\in}{2} + e^{-(p-n)\delta} \int_\delta^\infty |f(x)| e^{-nx} dx
\end{aligned}$$

Criterion (I) gives us that,

$$A = \int_\delta^\infty |f(x)| e^{-nx} dx \le \int_0^\infty |f(x)| e^{-nx} dx$$

Choose $p \ge n + \frac{1}{\delta} \ln\left(\frac{2A}{\in}\right)$, so that,

$$\begin{aligned}
|F(p)| &< \frac{\in}{2} + Ae^{-(p-n)\delta} \\
&\ge \frac{\in}{2} + Ae^{-\ln\left(\frac{2A}{\in}\right)} \\
&= \frac{\in}{2} + \frac{\in}{2} = \in
\end{aligned}$$

Since I can make $|F(p)|$ arbitrarily close to o for large p, I have $F(p) \to 0$ as $p \to \infty$.

Advantages of Laplace Transform

Laplace transforms methods offer the following advantages over the classical methods.

- It gives complete solution.
- Initial conditions are automatically considered in the transformed equations.

- Much less time is involved in solving differential equations.
- It gives systematic and routine solutions for differential equations.

Applications of Laplace Transforms

The Laplace transform is designed to analyze a specific class of time domain signals: impulse responses consisting of sinusoids and exponentials. The importance of this being that systems belonging to this class are extremely common in the sciences and engineering. The reason for this being they are quite often the solution to a differential equation as well as the fact that they are naturally occurring in the world. This makes it especially useful in signal processing, specifically analog signal processing, in which case the signal is continuous and is going to consist of a sinusoid or an exponential. The way in which analog signals are processed can be illustrated by the following diagram:

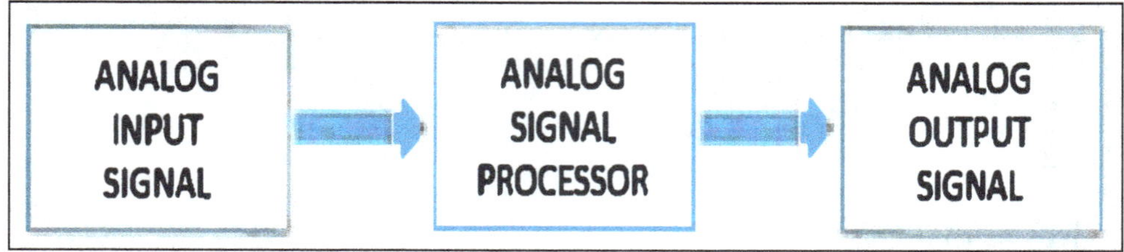

The system's input, processor and output are continuous time functions. For the input and output, the labels are $x(t)$ and $y(t)$ respectively. Also, for the purposes of this example we will label the analog signal processor in the middle by $h(t)$. Now, this is where the Laplace transform will finally come into play when doing analog signal processing. We will use the Laplace transform to figure out how the system behaves depending on what input is applied to it, and from there we can discover quite a few things about the system. This means we are trying to find out what the values of $y(t)$ are when we plug in $x(t)$ to the system. We can take the Laplace transform of this to get it into the complex s domain. By taking the Laplace transform, we get $X(s)$ and $Y(s)$, replacing our previous functions, $x(t)$ and $y(t)$, along with getting the transfer function, H(s). Note that $H(s)$ is the analog signal processor from the previous diagram and that the equation that will be mentioned below applies to many more fields than just analog signal processing. The reason we include it is because we take the Laplace transform of the processor as well so to get an accurate equation. It is also the processor that $X(s)$ goes through to give the output $X(s)$. This relationship can be seen in the following diagram, replacing the previous diagram with another one where the variables are now in the complex plane:

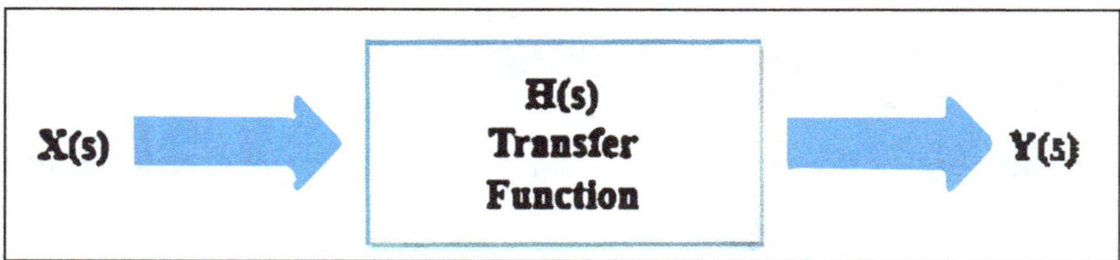

With this new system in the s plane, we can now figure out what the value of the transfer function,

$H(s)$, is. We do this by first writing the equation in the form we know we can write it as, by recognizing that to get $Y(s)$ we have to multiply the other two together.

$$Y(s) = H(s)X(s)$$

However, while knowing $Y(s)$ is useful, we truly want to know what H(s) is, so we just divide both sides by $X(s)$,

$$H(s) = \frac{Y(s)}{X(s)}$$

The importance of the equation directly above cannot be stressed enough when doing signal process Y' as well as many other fields where a transfer function is employed. By figuring out $\frac{Y(s)}{X(s)}$, we can find the transfer function of the system's value, giving us a lot of necessary information so we can then proceed to doing other work such as adjusting the filter or the signal to get the desired output wave. We can then use Laplace transforms to discover what $x(t)$ and $y(t)$ are, if we need to, these two being the original measures of the signal wave's input and output with respect to time. By doing this, we can gleam some information on what exactly we are working with if the value of original wave is one that we are unaware of. For now though, we are more focused on $H(s)$ and h(t) as these two give us much more information that is extremely crucial. The most important aspect of the equation giving us $H(s)$ is that by knowing what $H(s)$ is, we can discover if the system is stable. If it is, then we can discover what the frequency response of the system is, a rather important value to know. With the frequency response, we will know what our filter is doing and how to get the final result we are aiming for, as well as allowing us to adjust our sound waves to fix any issues with the filter if we need to. These pieces of information that are so vital to signal processing come from, as we have shown, the Laplace transform.

For a tangible example, we can see how exactly a filter works in reality by creating a sound wave and running it through a mathematical software capable of processing these signals. Using these types of software, the Laplace transform and all the resulting computation is done for us, making it much more convenient. It even includes the frequency response to discover how exactly to adjust the filter.

Example: Here are all of the diagrams of the process.

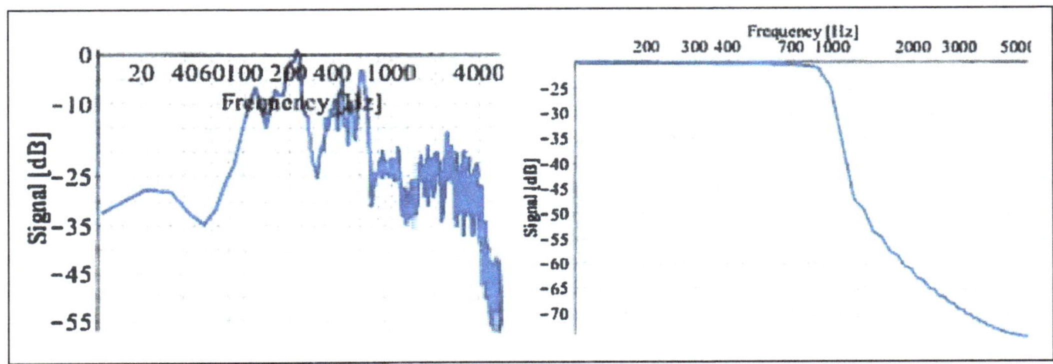

Original Input Signal and Frequency Response Respectively.

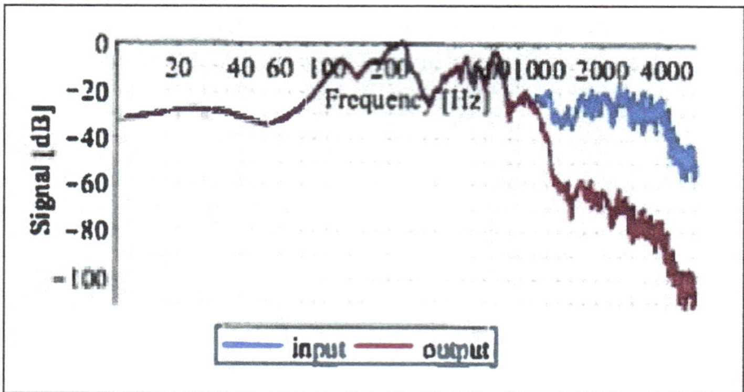

Input Wave vs. Output Wave.

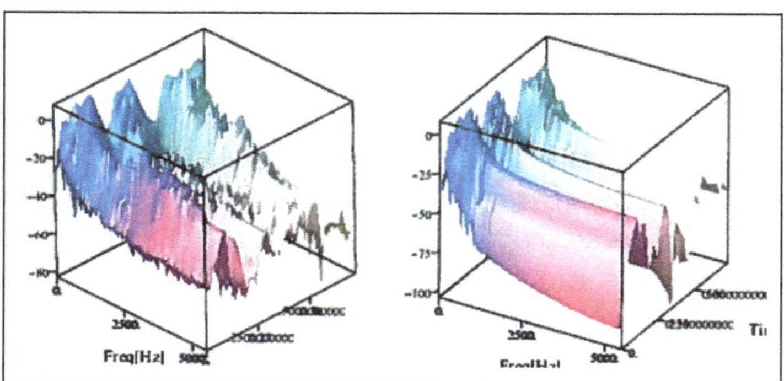

Frequency-Time Response for the Input Signal and Output Signal respectively.

The first diagram is the original signal before it has gone through any filter, with the second being its frequency response so we can alter the filter to work for the signal. The filter used in this example is an analog low pass filter with the cut off frequency boundaries being 1000 to 5000. We can see in the third diagram a comparison between the two filters with the blue wave being the input and the red wave being the output. However, this doesn't exactly give us a clear image of anything other than the frequency being lower, this is where the the last two diagrams come in. When comparing the two images, we can see that at the end of the signal, the curves become much smoother in the output signal, showcasing how much of an impact the filter truly had on the signal wave. While it is harder to see in graph form, the 3-D model does an excellent job of depicting the stark contrast between input and output.

Example: Let's look at another signal wave with a different type of filter, still utilizing the Laplace transform, in order to help showcase the versatility of it. To start, here are all the diagrams along with their labels:

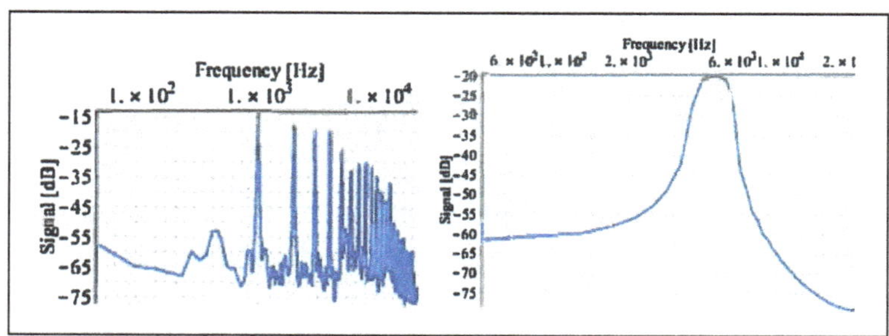

Original Input Signal and Frequency Response respectively

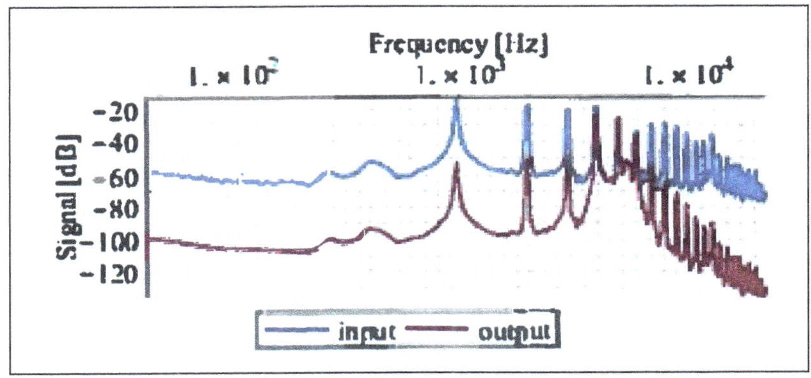

Input Wave vs. Output Wave.

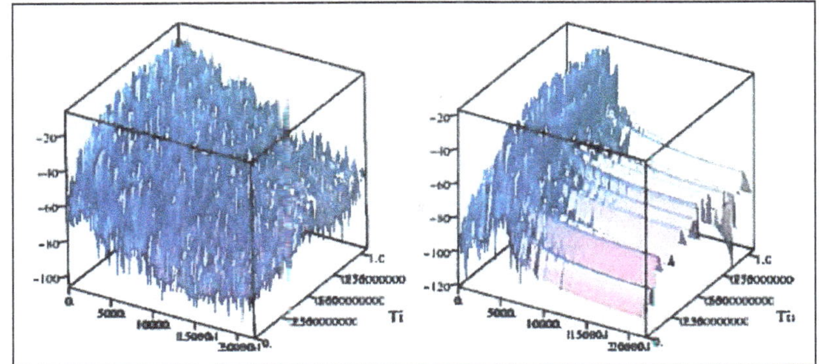
Frequency-Time Response for the Input Signal and Output Signal respectively.

The diagrams are in the same order as the previous example, with the first being the original input, the second being the frequency response, the third being the input vs. ouput and the last two being 3D images of the waves to showcase what truly is happening. We can see just how useful this filter is, the input wave is a complete mess while the output wave is almost completely smooth. The big difference in this example, versus the previous one, being that instead of a low pass filter, a Band-pass filter was used instead with the upper and lower boundaries being 4000 to 6000. An example of one of these is an RLC circuit. What this filter does is allow frequencies within a certain range to go through it whilst rejecting all other ones, as does the majority of other filters. The signifance of this one is that it is another filter that utilizies the Laplace transform, meaning it is an analog filter.

The process discussed above is an extremely significant process to the world today and as such is one of the major areas where the Laplace transform is used. It is important to note that this type of analysis, and subsequent processing of the signal is used in cellphones, a device that many would be unable to part with in modem society along with speakers, microphones and many other devices used by the general population.

Circuit Analysis

The Laplace transform actually gained its popularity from its use in analyzing electrical circuits due to Oliver Heaviside, an electrical engineer. By using Laplace transforms we can analyze an electrical circuit to discover its current, its maximum capacity and figure out if anything is wrong with the circuit. This is crucial for engineers, electrical engineers in particular, in doing their jobs to ensure the necessary machines and technology is working properly.

To start, let's show how this works in a simple RLC circuit. However, this does not mean it isn't used for more advanced types of circuits as well. For a visual aid, here is a diagram of a RLC circuit:

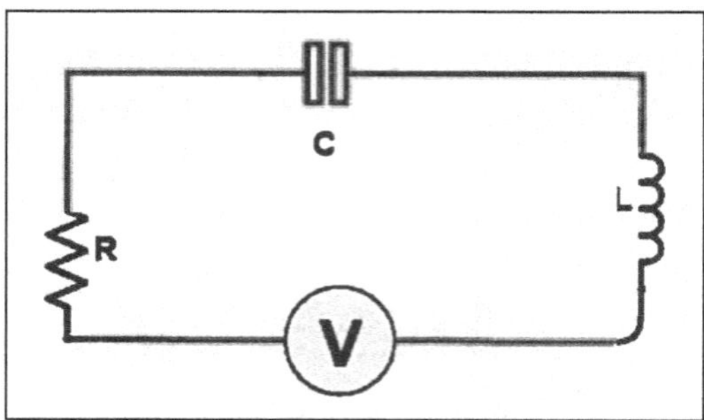

First let's identify the individual symbols on the circuit and what they mean. Also, while doing this it would help to identify what is used to measure each of these different pieces of the circuit for future reference. The symbols are as follows: R means resistor which is measured in ohms, L means the inductor which has inductance measured in henrys, C is the capacitor which has capacitance measured in farads and finally, V stands for the generator or battery and is measured in volts. Something to note is that another symbol commonly used for V is E when making diagrams of circuits. We can measure the charges of the capacitors and the currents by modeling them as functions of time. The equation that is used to model circuits and then subsequently used to analyze the circuits after solving it is as follows,

$$V(t) = RI + L'\frac{1}{C}Q$$

The remaining variable left to be defined is Q, which is normally the variable used to represent the charge of a circuit. We get this equation due to the fact that the voltage drop across a circuit is modeled by the following equations:

- The voltage drop across a resistor of a circuit is modeled by RI where $I = \frac{dQ}{dt}$.
- Across an inductor it is modeled by $L\frac{dI}{dt}$, and since we know $I = \frac{dQ}{dt}$, we simplify this to get $L\frac{d^2Q}{dt^2}$ which we can then reduce even further to LI'.
- Across a capacitor it is modeled by $\frac{1}{C}Q$.
- Across a generator it's modeled by $-V$.

By taking the Laplace transform of this equation, after plugging in values for the individual pieces of the circuit, and manipulating the resulting equation to take the inverse transform we can get a final solution to our circuit.

Before we go further, it is necessary to note that when we acquired the equation for $V(t)$, we actually used Kirchhoff's Laws. Due to the necessity of knowing these laws when doing circuit analysis, they are as follows:

- The algebraic sum of the currents flowing toward any junction point is equal to zero.

- The algebraic sum of the potential drops, or the voltage drops, around any closed loop is equal to zero.

The first of these two laws is often referred to as Kirchhoff's Current Law and the second of the two as Kirchhoff's Voltage Law. These two laws are extremely important to circuit analysis, as without them, the equation that we are using to model the circuit would not work. In some cases, only of the laws needs to be applied to get the equations. However, this is usually due to it being a rather simple circuit, such as the circuit in the first example.

Now that the circuit's components have been labeled we can showcase how exactly a Laplace transform is used in an introductory example followed by a more complex example.

Example:

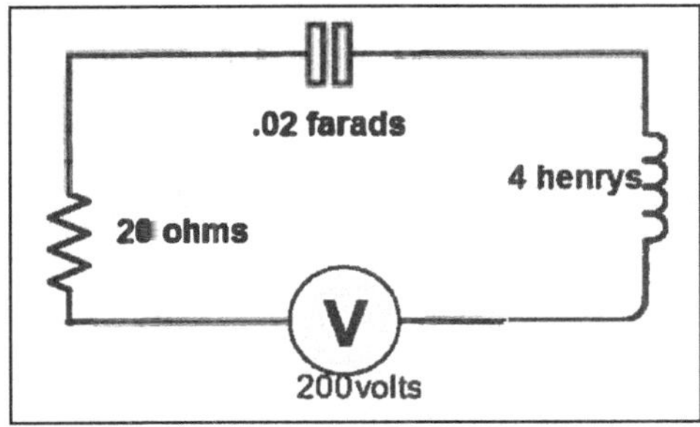

Based on the diagram above, our circuit has an inductor of 4 henrys, a resistor of 20 ohms and a capacitor of .02 farads. As for the charge and current, let's set a condition so that the charge on the capacitor, and current in the circuit, be 0 at t=0. Let's find the charge on the capacitor at any time t besides 0, where V is equal to 200 volts. So then we get the following,

$$4\frac{dI}{dt}+20I+\frac{1}{.02}Q=200$$

Since $I=\frac{dQ}{dt}$,

$$4\frac{d_2Q}{dt^2}+20\frac{dQ}{dt}+50Q=200$$

It is important to take into account that we have the following initial conditions due to our charge at t = 0 being 0.

- $Q(0)=0$,
- $Q'(0)=0$.

Now, we know the following is true:

- $\frac{d^2Q}{dt^2}=Q''$,
- $\frac{dQ}{dt}=Q$.

With this, we can rewrite the original equation,

$$Q''+5Q'+\frac{25}{2}Q=50$$

Now, we take the Laplace transform,

$$L\{Q''+5Q'+\frac{25}{2}Q\}=L\{50\}$$

$$=\{s^2q-sQ(0)-Q'(0)\}+5\{sq-Q(0)\}+\frac{25}{2}q=\frac{50}{s}$$

Recall our initial conditions to simplify this further,

$$q(s^2+5s+12.5)=\frac{50}{s}$$

$$=q=\frac{50}{s\left(s^2+5s+\frac{25}{2}\right)}$$

The goal is to take the inverse Laplace transform so that we can get the answer back in the original domain of time, but as of right now it isn't clear what function we get when taking the inverse transform. Since it isn't clear what the inverse transform function would be, we need to manipulate the equation. To start is partial fraction expansion of the equation, by doing this we get,

$$\frac{50}{s\left(s^2+5s+\frac{25}{2}\right)}=\frac{A}{s}+\frac{Bs+C}{s^2+5s+\frac{25}{2}}$$

So, by way of doing partial fraction expansion,

$$50=A\left(s^2+5s+\frac{25}{2}\right)+Bs^2+Cs$$

From here we solve for the individual variables. By plugging in O for s, we solve for A. Then if we plug that solution back in we can find B and C. By doing this, we end up with the following for the individual variables:

- $A=4$,
- $B=-4$,
- $C=-20$.

Now, we just plug these back into the original equation,

$$\frac{4}{s}-\frac{-4s-20}{s^2+5s+\frac{25}{2}}$$

From here we manipulate the equation to fit one in the form from the table.

$$= \frac{4}{} - 4\frac{-}{\left(s+\frac{5}{2}\right)^2 + \frac{25}{2}} - 10\frac{1}{\left(s+\frac{5}{2}\right)^2 + \frac{25}{2}}$$

With the equation now fitting the table on Laplace transforms, we can take the inverse transform.

$$L^{-1}\{\frac{4}{s} - 4\frac{s+\frac{5}{2}}{\left(s+\frac{5}{2}\right)^2 + \frac{25}{4}} - 10\frac{1}{\left(s+\frac{5}{2}\right)^2 + \frac{25}{4}}\}$$

$$= 4 - 4e^{-\frac{5}{2}t}\cos(\frac{5}{2}t) - 4e^{-\frac{5}{2}t}\sin(\frac{5}{2}t)$$

So our charge at any time, t, t > 0 is the equation above. We can see what this looks like in the form of a graph via Maple to determine exactly what it is.

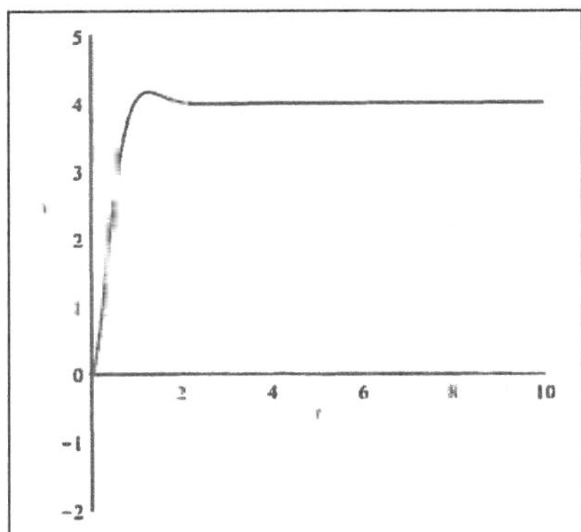

We can see from this graph that the charge maxes out a little after 4C and then flattens out at 4C.

After doing an example of a rather basic circuit, let's do one a little more advanced with multiple loops to showcase how the Laplace transform is utilized in a more advanced case.

Example:

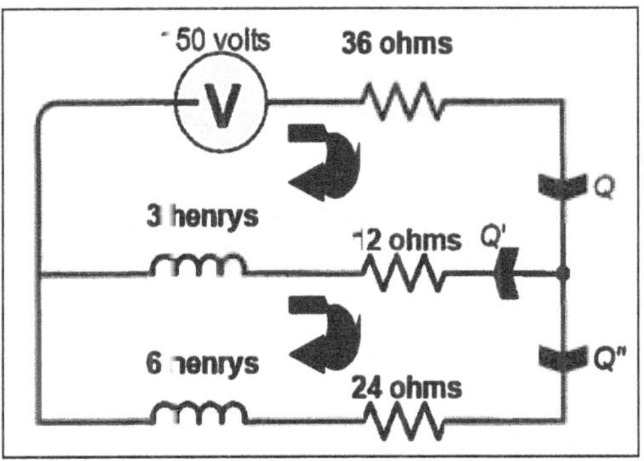

We have a circuit with two different branches, let's figure out what the currents are in each of these branches when the initial is zero. Due to Kirchhoff's second law, we know that the sum of the voltage on a closed loop is zero, and we can see that our loops are closed from the diagram above. Let's make Q the current around the top part of the circuit, and then let Q' and Q" be the respective currents that divide at the junction point so that $Q = Q' + Q''$. Also, it is important to note that we have the following intitial conditions:

- $Q(0) = 0$,
- $Q'(0) = 0$.

Now that we know the initial conditions, let's analyze this circuit. To do this we need to apply Kirchhoff's second law to these two loops to get the following equations,

- $-12Q' - 3\frac{dQ'}{dt} + 6\frac{dQ''}{dt} + 24Q'' = 0$,
- $36Q + 3\frac{dQ'}{dt} + 12Q' = 150$.

To make things easier, let's work with the first equation.

- $-4Q' - \frac{dQ'}{dt} + 2\frac{dQ''}{dt} + 8Q'' = 0$,
- $12Q + 3\frac{dQ'}{dt} + 4Q' = 150$.

To make things easier, let's work with the first equation.

$$\begin{aligned} &L\{-4Q' - \frac{dQ'}{dt} + 2\frac{dQ''}{dt} + 8Q''\} = 0 \\ &-4q' - (sq' - Q'(0)) + 2(sq'' - Q(0)) + 8q'' \\ &\qquad = 4q' - sq' + 2sq + 8q \\ &\qquad = (s+4)q' - (2s+8)q'' \end{aligned}$$

It follows that,

$$\begin{aligned} &(s+4)q' = (2s+8)q \\ &\quad = q' = \frac{2s+8}{s+4}q \\ &\quad = q' = 2q'' \end{aligned}$$

Now that we know what q' is, let's focus on the second equation. If we apply Kirchhoff's seocnd law, we can alter it to get the following:

$$\frac{dQ'}{dt} + 8Q' + 6Q'' = 50$$

The reason for doing this is so we can take the Laplace transform.

$$L\{\frac{dQ'}{dt}+8Q'+6Q"\}=L\{50\}$$

$$=\{sq'Q'(0))+8q'+6q'=\frac{50}{s}$$

Recall our initial condition to simply the equation,

$$(s+8)q'+6q"=\frac{50}{s}$$

Since we know that $q'=2q"$, we are going to substitute it in.

$$(s+8)q"+6q"=\frac{50}{s}$$

$$=q"(10s+14)=\frac{50}{s}$$

$$=q"=\frac{50}{s(10s+14)}$$

Since q" is now by itself, we can take the inverse Laplace transform of the equation,

$$L^{-1}\{q"\}=L^{-1}\{\frac{50}{s(10s+14)}\}$$

$$Q"=\frac{25}{7}-2e^{\frac{7}{5}t}\frac{25}{7}$$

It follows that since $q'$is double this,

$$Q'=\frac{50}{7}-2e^{\frac{7}{5}t}\frac{25}{7}$$

Recall the previous equation, $Q=Q'+Q"$. From this, it follows that,

$$Q=\frac{25}{7}-e^{\frac{7}{5}t}\frac{25}{7}+\frac{50}{7}-2e^{\frac{7}{5}t}\frac{25}{7}$$

$$Q=\frac{75}{7}-3e^{\frac{7}{5}t}\frac{25}{7}$$

At any time $t, t>0$ the circuit's current will have the value denoted by the equation given by Q. We can see this via a graph to truly understand what this means.

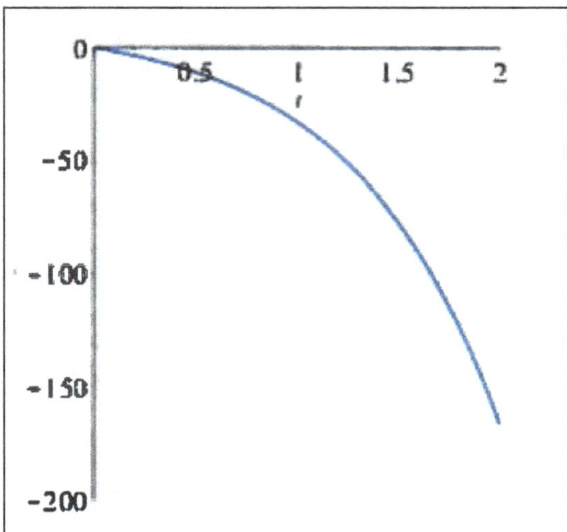

We can see from this graph that the charge is exponentially decreasing and will continue to drop.

The process of analyzing circuits is used by engineers who wish to gain a better understanding of the circuit they are currently working with. While the two examples are not increasingly complex, lacking a switch for both, the same principle is applied to any RLC circuit. By using a Laplace transform for circuit analysis, we get the automatic inclusion of the initial conditions, in the examples case these were $Q(0)=0$ and $Q'(0)=0$, giving us an entirely complete solution of the analysis. The fact that we get our initial conditions automatically included in the solution is arguably the main reason why the transform gained such popularity in doing circuit analysis.

Comparison between Fourier and Laplace Transform

First, lets take a look at the Fourier Transform (FT) of a CT signal, $x(t)$,

$$X(\omega)=F\left[x(t)\right] \;=x(t)^{e-j\omega t}\;\; dt.$$

The FT transforms a time-domain signal into a frequency-domain signal, telling us how the signal's energy (−information) is spread across sinusoids of different frequencies. This property makes the FT invaluable in many signal processing tasks such as audio engineering and wireless communications (though, the applications of the FT are limitless). Now, remember that the FT had the following condition: the $F\left[x(t)\right]$ only exists if

$$\int\left|x(t)\right|dt<\infty.$$

$$-\infty$$

In other words, $F\left[x(t)\right]$ exists only if the total energy of our signal is bounded, that is, the area underneath its curve is finite.

Why does the FT have this condition? Lets attempt to find the FT of the two unacceptable signals shown above. For the first case, we observe a unit step function,

$$x(t)=u(t),$$

which is defined to have a value of 1 for all positive t. So, attempting to calculate $F\left[u(t)\right]$,

$$X(\omega)=F(u(t))=\int_{-\infty}^{\infty}u(t)^{e-j\omega t}\,dt,$$

The integral ∞ is, technically, an improper integral. We will note that by setting a bound to infinity we are actually saying R that we will calculate this integral as the upper end approaches infinity in the limit. Remembering this fact we find

$$X(\omega)=\int_{-\infty}^{\infty}u(t)^{e-j\omega t}\,dt,$$

Unfortunately for us, the term $LIM_{t\to\infty}e^{-jt}$ doesn't converge to anything in the limit since the term,

$$e^{-j\omega t}={}^{1}{}_{2}\,[COS\,\omega t-j\,SIN\,\omega t]$$

and here, again, we see that for our stated condition $\alpha>0$, the integral diverges as the term $LIM_{t\to\infty}e^{\alpha t}e^{-j\omega t}$ goes to infinity. And so, $X(\omega)$ does not exist. While the FT is useful for analyzing signals in terms of their frequency composition, many kinds of signals, such as the two we have just investigated, defy strict Fourier transformation (though general transforms can be arrived at.)

To analyze a more general set of functions, including functions which may not have FTs, we can use the Laplace transform (LT). The LT can be thought of as a generalized FT. The LT is defined as,

$$X(\omega)=F(u(t))=\int_{-\infty}^{\infty}u(t)e^{-j\omega t}dt,$$

where,

$$s=\sigma+j\omega.$$

where the FT of a signal, $X(\omega)$, was a function of only imaginary numbers, the LT of the same signal, X(s), is a function of complex numbers consisting for both real and imaginary components. We can view the FT as a special case of the LT for which $\sigma=0$.

The inverse FT tells us that the function $x(t)$ can be constructed from complex sinusoids which are denoted by the $e^{j\omega t}$ term. An example of one such complex sinusoid is given in figure. These complex sinusoids have a fixed amplitude across time. It is this fixed amplitude which prevents us from analyzing signals whose energy is not bounded. Simply put, the FT's complex sinusoids do not have enough explanatory power for these functions.

Now, let us contrast the FT's construction of $x(t)$ with that of the LT. The inverse LT is defined as,

$$x(t)=\frac{1}{2\pi}\int_{\sigma-j\infty}^{\sigma+j\infty}X(\omega)e^{(\sigma+j\omega)}\,dt$$

Rather than complex sinusoids of fixed amplitude, the LT defines $x(t)$ in terms of complex sinusoids whose amplitudes exponentially grow or decay with respect to time, figure gives an example of one such exponentially growing sinusoid. The additional exponential term est gives the LT much more flexibility than the FT for representing signals whose energy is unbounded.

Now, lets use the LT to find the transformation of the unit step and exponential growth functions for which the FT does not exist. Note that we are using the unilateral LT, which integrates over $\left[-\infty,\infty\right]$, rather than the bilateral LT which integrates over [−∞, ∞]. We use the unilateral LT for analyzing casual signals, i.e. signals for which $x(t)=0, t<0$. In our case, both the unit step and exponential growth functions we analyze here are causal. First, the unit step, Here we see the extra leverage that the LT gives us over the FT. We now have an additional term, σ which we can use to force this integral to converge. Specifically,

$$\underset{t\to\infty}{\text{LIM}}\, e^{-\sigma t}\, e^{-j\omega t} = 0, \;\; \sigma > 0.$$

However, that this convergence only occurs for certain values of σ, in this case $\sigma > 0$. Therefore,

$$X(s) = L[u(t)] = \frac{1}{\sigma + J\omega} \overset{[-1]}{=} \frac{1}{s}, \sigma > 0$$

We make the note here that the region in which the LT exists (in this case, $\sigma > 0$) is called the Region of Convergence (ROC) for the LT. Each different LT has a different ROC in which the LT exists. Finally, we see that the step function, which did not have a FT due to its infinite energy, does indeed have an LT (though only within the ROC).

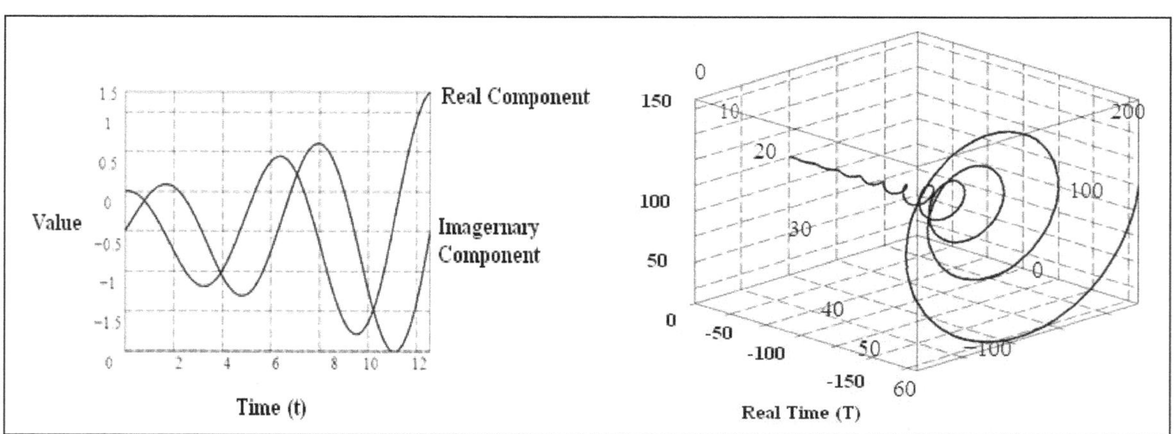

Complex sinusoid $e^{(0.1-j)t}$.

Region of Convergence

Whether the Laplace transform X(s) of a signal $x(t)$ exists or not depends on the complex variable as well as the signal itself. All complex values of for which the integral in the definition converges form a region of convergence (ROC) in the s-plane. X(s) exists if and only if the argument is inside the ROC. As the imaginary part $\omega = \text{Im}\left[s\right]$ of the complex variable $s = \sigma + j\omega$ has no effect in terms of the convergence, the ROC is determined solely by the real part $\sigma = \text{Re}\left[s\right]$.

Example: The Laplace transform of $x(t) = e^{-at}u(t)$

$$X(s) = \mathcal{L}[x(t)] = \int_0^{\infty} e^{-at}e^{-st}\,dt = \int_0^{\infty} e^{-at}e^{-(\sigma+j\omega)t}\,dt$$

$$= -\frac{1}{a+\sigma+j\omega}e^{-(a+\sigma+j\omega)t}\Big|_0^{\infty}$$

For this integral to converge, we need to have,

$$a+\sigma > 0 \quad \text{or} \quad \sigma = \text{Re}[s] > -a$$

and the Laplace transform is,

$$X(s) = \frac{1}{(\sigma+a)+j\omega} = \frac{1}{s+a}$$

As a special case where $a = 0$, $x(t) = u(t)$ and we have,

$$\mathcal{L}[u(t)] = \frac{1}{s}, \quad \sigma = \text{Re}[s] > 0$$

Example: The Laplace transform of a signal $x(t) = -e^{-at}u(-t)$ is,

$$X(s) = -\int_{-\infty}^{0} e^{-at}e^{-st}\,dt = -\int_{-\infty}^{0} e^{-(a-\sigma+j\omega)t}\,dt = \frac{1}{a+\sigma+j\omega}e^{-(a+\sigma+j\omega)t}\Big|_{-\infty}^{0}$$

Only when $a+\sigma < 0$ or $\sigma = \text{Re}[s] < -a$

will the integral converge, and Laplace transform $X(s)$,

$$X(s) = \frac{1}{a+\sigma+j\omega} = \frac{1}{a+s}$$

Again as a special case when $a = 0$, $x(t) = -u(-t)$ we have,

$$\mathcal{L}[-u(-t)] = \frac{1}{s}, \quad \sigma = \text{Re}[s] < 0$$

Comparing the two examples above we see that two different signals may have identical Laplace transform $X(s)$, but different ROC. In the first case above, the ROC is $\text{Re}[s] > 0$, and in the second case, the ROC is $\text{Re}[s] < 0$. To determine the time signal $x(t)$ by the inverse Laplace transform, we need the ROC as well as $X(s)$. Now we turn our attention to the exponential growth function, And, once again, we see that the LT gives us to ability to force this integral to converge by setting $\sigma - \alpha > 0$, that is, $\sigma > \alpha$.

Before we move on, let us make a few observations. First, we see that the LT of the unit step function and of the exponential growth function are related. Let us, instead of considering just an exponentially growing function, allow α to take on any value, positive or negative. We now have a function which:

- Grows when α > 0,
- Is constant when α = 0,
- Decays when α < 0.

When we set $\alpha = 0$, we see that $e^{\alpha t}u(t) \rightarrow u(t)$. Thus,

$$L[e^{(\alpha=0)t}u(t)] = \frac{1}{s-0} = 1/s = L[u(t)].$$

Now, lets look at the ROC for the LT of our general, causal, exponential function, $x(t) = e^{at}\, u(t)$. The ROC for differing values of α are given in figs. In the case that $\alpha = 0$ or $\alpha > 0$, i.e. a unit step and an exponential growth function, the FT does not exist.

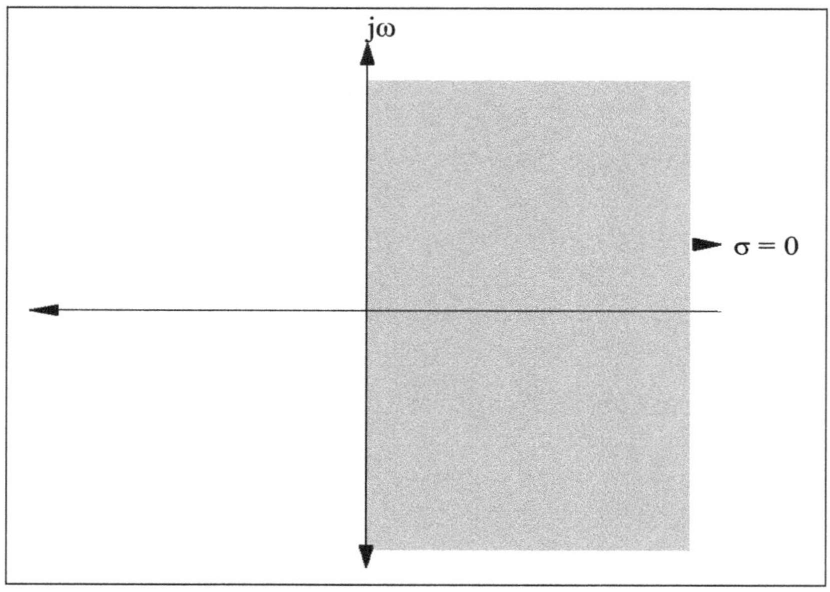

ROC on the complex plane for $L[e^{at}u(t)]$ for $\alpha = 0$.

It does not contain the $j\omega$ axis. However, when $\alpha < 0$, a FT for the exponential function does exist, and the $j\omega$ axis is contained within the ROC of the LT. This makes sense, because the FT is a special case of the LT when $\sigma = 0$.

Specifically, if $x(t)$ is causal and its FT exists (in the strict sense), then,

$$F[x(t)] = X(s)\big|_{s=j\omega}$$

where $X(s) = L[x(t)]$. The existence of $F[x(t)]$ is equivalent to having the $j\omega$ axis of the complex plane within the ROC of $L[x(t)]$. If $j\omega$ is not within the ROC, then,

$$L[x(t)]\Big|_{s=j\omega} = 6\,F[x(t)]$$

For example, lets look at the unit step,

- $F[u(t)] =_{j\omega}{}^{1} + \pi\delta(\omega)$,
- $L[u(t)] = \frac{1}{s}$

Here, we see that $L[u(t)]6 = F[u(t)]$ because of the ROC of $L[u(t)]$ does not contain the jì axis.

And so, we see that the Laplace transform and the Fourier transform are linked together, with the Fourier transform being a special case of the Laplace transform. The Laplace transform exhibits greater explanatory power than the Fourier transform as it allows for the transformation of functions with unbounded energy.

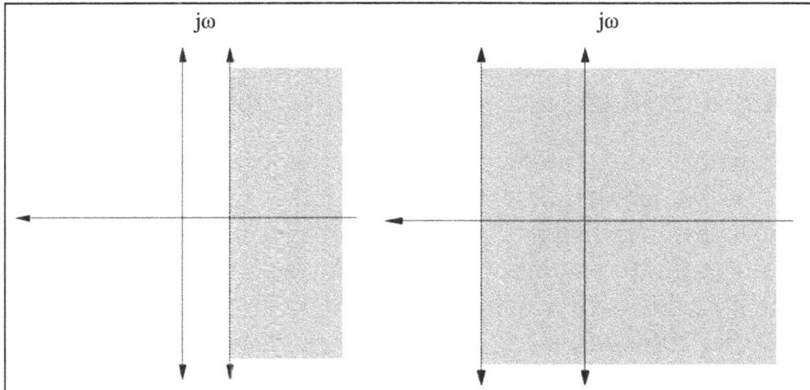

ROC on the complex plane for $L[e^{at}u(t)]$ for $\alpha > 0$.

$\sigma\alpha < 0$

Difference between Fourier and Laplace Transformation

Laplace Transform does a real transformation on complex data but Fourier Transform does a complex transformation on real data. it means:

$$\begin{aligned}
\mathcal{L}\{f(t)\} &= \int_{0-}^{\infty} e^{-st} f(t)dt \\
&= \left[\frac{f(t)e^{-st}}{-s}\right]_{0-}^{\infty} - \int_{0-}^{\infty} \frac{e^{-st}}{-s} f'(t)dt \quad \text{(by parts)} \\
&= \left[-\frac{f(0-)}{-s}\right] + \frac{1}{s}\mathcal{L}\{f'(t)\},
\end{aligned}$$

At the presence of some condition you can take the Fourier transformation by placing j instead of s in Laplace transform. To use Laplace to find an output given a system and input, you find the Laplace of the input $X(s)$, and the system $H(s)$, multiply them together to find the output $Y(s)$, then inverse transform that to find $Y(t)$. Use Fourier Transforms you split the input signal $X(s)$ into many pure sinusoids each of a known amplitude, phase, and frequency w, then directly find the associated output sinusoid for each of the inputs (same as the input but with a gain of $|H(jw)|$ and with an added phase of $\text{angle}(H(jw))$).

Fourier transformation sometimes has physical interpretation, for example for some mechanical models where we have quasi-periodic solutions (usually because of symmetry of the system) Fourier transformations gives You normal modes of oscillations. Sometimes even for nonlinear system, couplings between such oscillations are weak so nonlinearity may be approximated by power series in Fourier space. Many systems has discrete spatial symmetry (crystals) then solutions of equations has to be periodic so FT is quite natural (for example in Quantum mechanics).

With any of normal modes You may tie finite energy, sometimes momentum etc. invariants of motion. So during evolution, for linear system, such modes do not couple each other, and system in one of this state leaves in it forever. Every linear physical system has its spectrum of normal modes, and if coupled with some external random source of energy (white noise), its evolution runs through such states from the lowest possible energy to the greatest.

When you Laplace transform the system, you will get the final system response, if you know the initial conditions of the system. These conditions are at t =0, can be easily obtained from the equation. The Fourier transform helps in analyzing the system response in a way different from Laplace. It breaks the signal into a number of sine and cosine waves (actually infinite), where you can have an insight to how the system is behaving by observing the amplitudes of each of the sine and cosine waves. Fourier transforms are for converting/representing a time-varying function in the frequency domain.

A laplace transform are for converting/representing a time-varying function in the "integral domain" The Laplace and Fourier transforms are continuous (integral) transforms of continuous functions. The Laplace transform maps a function $f(t)$ to a function $F(s)$ of the complex variable s, where $s = \sigma + j\omega$.

Since the derivative $f(t)=df(t)\,dt$ maps to $sf(s)$, the Laplace transform of a linear differential equation is an algebraic equation. Thus, the Laplace transform is useful for, among other things, solving linear differential equations.

If we set the real part of the complex variable s to zero, $\sigma = 0$, the result is the Fourier transform $F(j\omega)$ which is essentially the frequency domain representation of $f(t)$.

The Z transform is essentially a discrete version of the Laplace transform and, thus, can be useful in solving difference equations, the discrete version of differential equations. The Z transform maps a sequence $f(n)$ to a continuous function $F(z)$ of the complex variable $z=rej\Omega$.

If we set the magnitude of z to unity, r=1, the result is the Discrete Time Fourier Transform (DTFT) $F(j\Omega)$ which is essentially the frequency domain representation of $f[n]$.

Fourier transform is defined only for absolutely integrable functions. Laplace transform is a generalisation to include all functions.

References

- LaplaceTransform: mathworld.wolfram.com, Retrieved 12 May, 2019
- "Computer explorations in Signals and Systems using MATLAB",Prentice Hall Signal Processing Series
- Understanding Digital Signal Processing, Second Edition Prentice Hall PTR, 2004. http:/ /flylib.com/books/ en/2. 729.1.67 /1
- Fourier and Laplace transforms. Cambridge University Press (2003)
- Dynamic analysis of functionally graded plates using the hybrid Fourier-Laplace transform under thermomechanical loading. Meccanica, 46(6), 1373-1392

6

Z-Transform

In signal processing, Z-transformation is used to transform a series of real or complex values into a complex frequency domain representation. Region of convergence, properties of Z transform, pole zero plot, inverse Z transform, etc. are some of the principles associated with it. This chapter discusses these principles related to Z Transform in detail.

A special feature of the z-transform is that for the signals and system of interest to us, all of the analysis will be in terms of ratios of polynomials. Working with these polynomials is relatively straight forward.

- Given a finite length signal $x[n]$, the z -transform is defined as:

  $$X(z) = \sum_{k=0}^{N} x[k] z^{-k} = \sum_{k=0}^{N} x[k] (z^{-k})^{k}$$

  where the sequence support interval is $[0, N]$, and z is any complex number.

- This transformation produces a new representation of $x[n]$ denoted $X(z)$.
- Returning to the original sequence (inverse z-transform) $x[n]$ requires finding the coefficient associated with the nth power of z^{-1}.
- Formally transforming from the time/sequence/n-domain to the z -domain is represented as:

  $$n-\text{Domain} \overset{z}{\leftrightarrow} z-\text{Domain}$$

  $$x[n] = \sum_{k=0}^{N} x[k]\delta[n-k] \overset{z}{\leftrightarrow} X(z) = \sum_{k=0}^{N} x[k] z^{-k}$$

- A sequence and its z-transform are said to form a z -transform pair and are denoted:

  $$x[n] \overset{z}{\leftrightarrow} X(z)$$

  - In the sequence or n-domain the independent variable is n.
  - In the z -domain the independent variable is z .

Example: $x[n] = \delta[n - n_0]$

Using the definition:

$$X(z) = \sum_{k=0}^{N} x[k] z^{-k} = \sum_{k=0}^{N} \delta[k - n_0] z^{-k} = z^{-n_0}$$

Thus,

$$\delta[n - n_0] \overset{z}{\leftrightarrow} z^{-n_0}$$

Example: $x[n] = 2\delta[n] + 3\delta[n-1] + 5\delta[n-2] + 2\delta[n-3]$

By inspection we find that:

$$X(z) = 2 + 3z^{-1} + 5z^{-2} + 2z^{-3}$$

Example: $X(z) = 4 - 5z^{-2} + z^{-3} + 2z^{-4}$

By inspection we find that:

$$x[n] = 4\delta[n] + 5\delta[n-2] + \delta[n-3] - 2\delta[n-4]$$

What can we do with the z-transform that is useful?

Region of Convergence

With the z-transform, the s-plane represents a set of signals (complex exponentials). For any given LTI system, some of these signals may cause the output of the system to converge, while others cause the output to diverge ("blow up").

The set of signals that cause the system's output to converge lie in the region of convergence (ROC).

The region of convergence, known as the ROC, is important to understand because it defines the region where the z-transform exists. The z-transform of a sequence is defined as:

$$X(z) = \sum_{n=-\infty}^{\infty} x[n] z^{-n}$$

The ROC for a given $x[n]$, is defined as the range of z for which the z-transform converges. Since the z-transform is a power series, it converges when $x[n]\ z^{-n}$ is absolutely summable.

Stated differently,

$$\sum_{n=-\infty}^{\infty} \left| x[n] z^{-n} \right| < \infty$$

must be satisfied for convergence.

Properties of the Region of Convergencec

The Region of Convergence has a number of properties that are dependent on the characteristics of the signal, $x[n]$.

The ROC cannot contain any poles. By definition a pole is a where $X(z)$ is infinite. Since $X(z)$ must be finite for all z for convergence, there cannot be a pole in the ROC.

If $x[n]$ is a finite-duration sequence, then the ROC is the entire z-plane, except possibly $z = 0$ or $|z| = \infty$. A finite-duration sequence is a sequence that is nonzero in a finite interval $n_1 \le n \le n_2$. As long as each value of $x[n]$ is finite then the sequence will be absolutely summable. When $n_2 > 0$ there will be a z^{-1} term and thus the ROC will not include $z = 0$. When $n_1 < 0$ then the sum will be infinite and thus the ROC will not include $|z| = \infty$. On the other hand, when $n_2 \le 0$ then the ROC will include z=0, and when n1≥0 the ROC will include $|z| = \infty$. With these constraints, the only signal, then, whose ROC is the entire z-plane is $x[n] = c\delta[n]$.

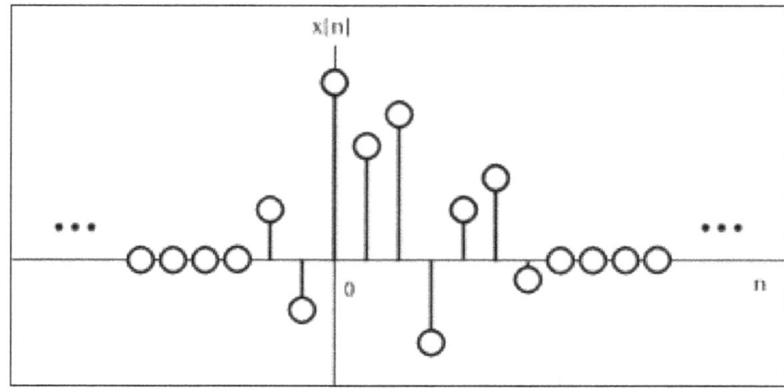

An example of a finite duration sequence.

The next properties apply to infinite duration sequences. As noted above, the z-transform converges when $|X(z)| < \infty$. So we can write:

$$|X(z)| = \left| \sum_{n=-\infty}^{\infty} x[n]z^{-n} \right| \le \sum_{n=-\infty}^{\infty} \left| x[n]z^{-n} \right| = \sum_{n=-\infty}^{\infty} |x[n]| (|z|)^{-n}$$

We can then split the infinite sum into positive-time and negative-time portions. So,

$$|X(z)| \le N(z) + P(z)$$

where,

$$N(z) = \sum_{n=-\infty}^{-1} |x[n]| (|z|)^{-n}$$

and

$$P(z) = \sum_{n=0}^{\infty} |x[n]| (|z|)^{-n}$$

In order for $|X(z)|$ to be finite, $|x[n]|$ must be bounded. Let us then set,

$$|x(n)| \leq C_1 r_1^n$$

For,

$$n \geq 0$$

From this some further properties can be derived:

If $x(n)$ is a right-sided sequence, then the ROC extends outward from the outermost pole in $X(z)$. A right-sided sequence is a sequence where $|x[n]=0|$ for $n < n_1 < \infty$. Looking at the positive-time portion from the above derivation, it follows that,

$$P(z) \leq C_2 \sum_{n=0}^{\infty} r_2^n \left(|z|\right)^{-n} = C_2 \sum_{n=0}^{\infty} \left(\frac{r_2}{z|}\right)^n$$

Thus in order for this sum to converge, $|z| > r_2$, and therefore the ROC of a right-sided sequence is of the form $|z| > r_2$.

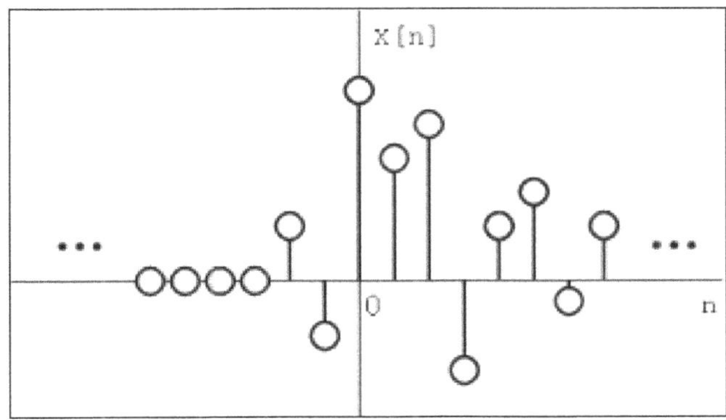

A right-sided sequence.

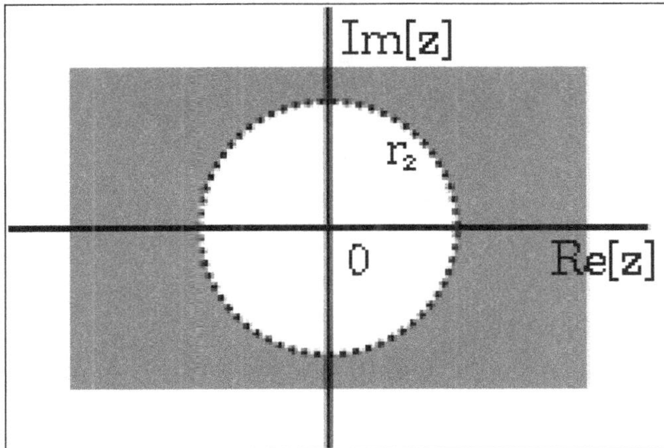

The ROC of a right-sided sequence.

If $x[n]$ is a left-sided sequence, then the ROC extends inward from the innermost pole in $X(z)$

a left-sided sequence is a sequence where $x[n]=0$ for $n>n_2>-\infty$. Looking at the negative-time portion from the above derivation, it follows that,

$$N(z)\le C_1\sum_{n=-\infty}^{-1} r_1^n\left(|z|\right)^{-n}=C_1\sum_{n=-\infty}^{-1}\left(\frac{r_1}{|z|}\right)^n = C_1\sum_{k=1}^{\infty}\left(\frac{|z|}{r_1}\right)^k$$

Thus in order for this sum to converge, $|z|<r_1$, and therefore the ROC of a left-sided sequence is of the form $|z|<r_1$.

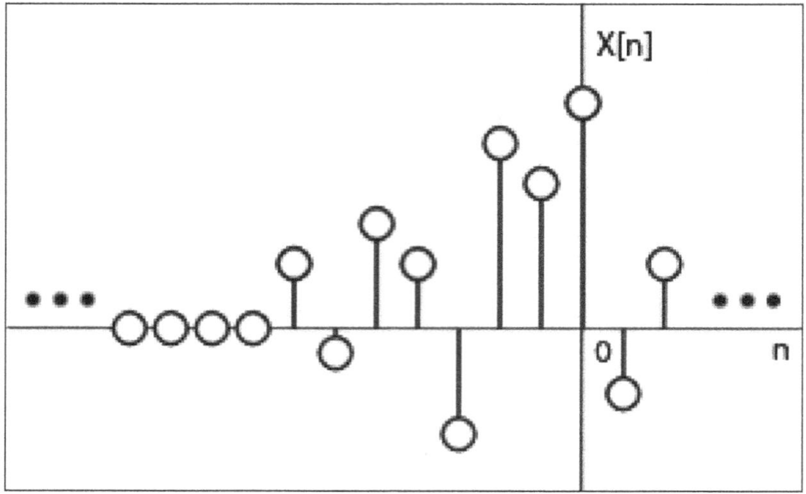

A left-sided sequence.

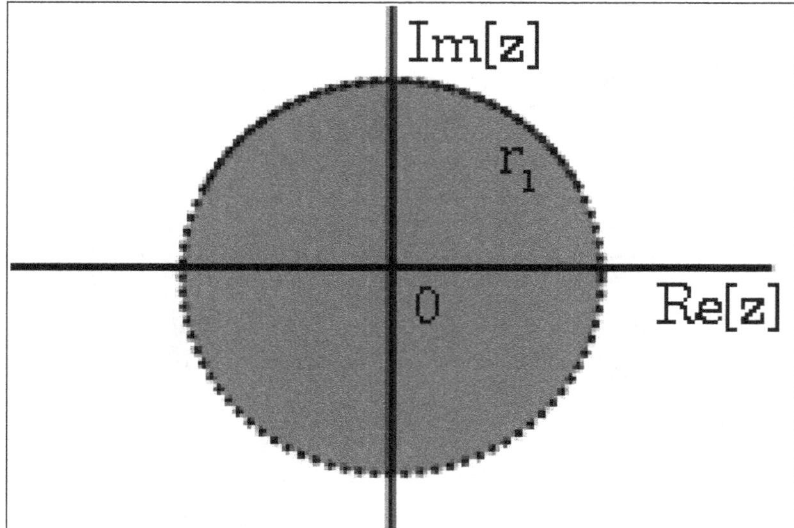

The ROC of a left-sided sequence.

If $x[n]$ is a two-sided sequence, the ROC will be a ring in the z-plane that is bounded on the interior and exterior by a pole. A two-sided sequence is an sequence with infinite duration in the positive and negative directions. From the derivation of the above two properties, it follows that if $-r_2<|z|<r_2$ converges, then both the positive-time and negative-time portions converge and thus $X(z)$ converges as well. Therefore the ROC of a two-sided sequence is of the form $-r_2<|z|<r_2$.

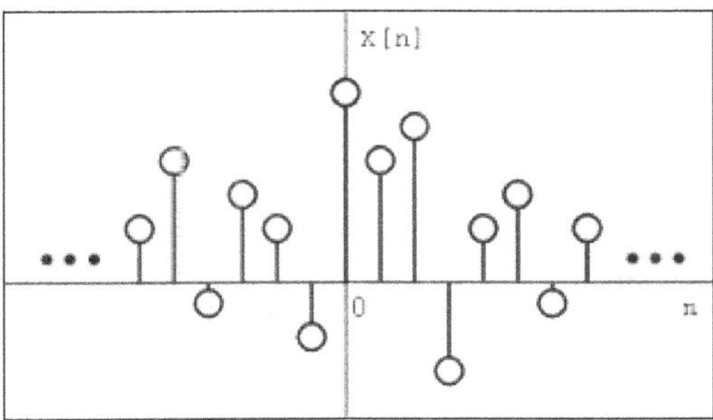

A two-sided sequence.

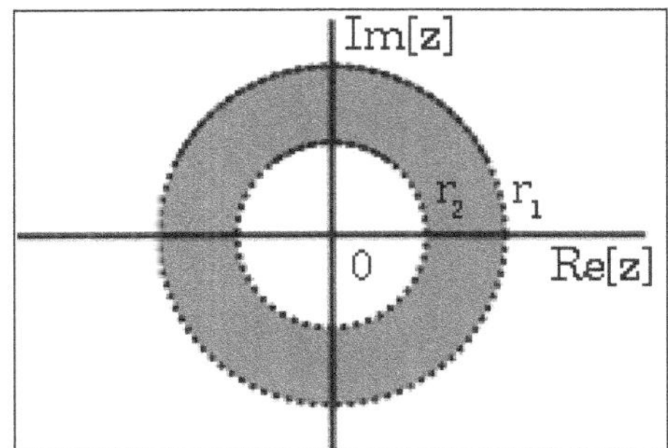

The ROC of a two-sided sequence.

Example:

Lets take,

$$x_1[n] = \left(\frac{1}{2}\right)^n u[n] + \left(\frac{1}{4}\right)^n u[n]$$

The z-transform of $\left(\frac{1}{2}\right)^n u[n]$ is $\frac{z}{z-\frac{1}{2}}$ with an ROC at $|z| > \frac{1}{2}$.

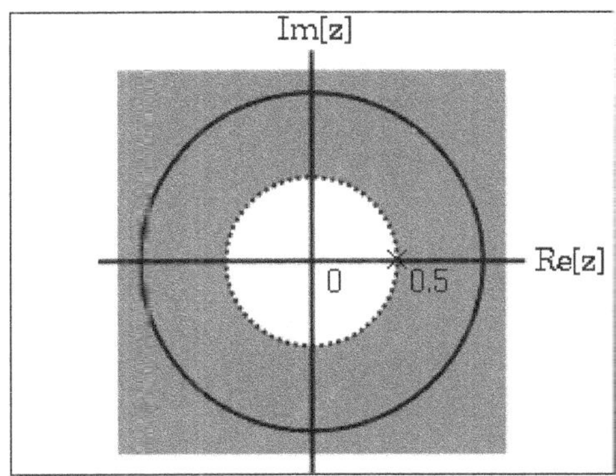

The ROC of $\left(\frac{1}{2}\right)^n u[n]$.

The z-transform of $\left(\frac{-1}{4}\right)^n u[n]$ is $\frac{z}{z+\frac{1}{4}}$ with an ROC at $|z| > \frac{-1}{4}$.

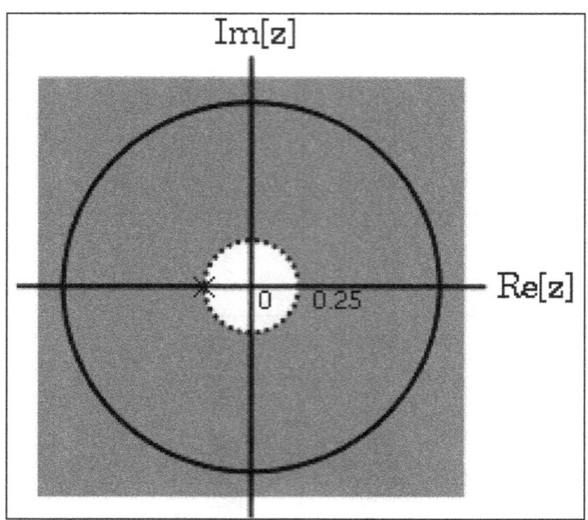

The ROC of $\left(\frac{-1}{4}\right)^n u[n]$.

Due to linearity,

$$X_1[z] = \frac{z}{z-\frac{1}{2}} + \frac{z}{z+\frac{1}{4}}$$
$$= \frac{2z\left(z-\frac{1}{8}\right)}{\left(z-\frac{1}{2}\right)\left(z+\frac{1}{4}\right)}$$

By observation it is clear that there are two zeros, at 0 and $\frac{1}{16}$, and two poles, at $\frac{1}{2}$, and $\frac{-1}{4}$. Following the above properties, the ROC is $|z| < \frac{1}{2}$.

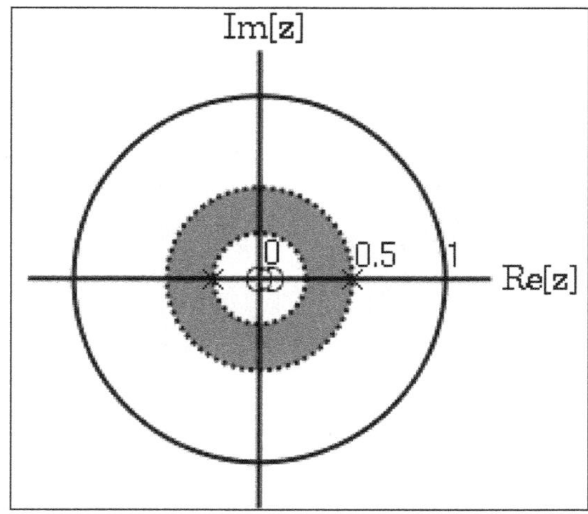

The ROC of $x_2[n] = \left(\frac{-1}{4}\right)^n u[n] - \left(\frac{1}{2}\right)^n u[(-n)-1]$.

Properties of Z-Transfrom

The z-transform has a set of properties in parallel with that of the Fourier transform (and Laplace transform). The difference is that we need to pay special attention to the ROCs. In the following, we always assume,

$$Z[x[n]] = X(z) \quad ROC = R_x$$

and

$$Z[y[n]] = Y(z) \quad ROC = R_y$$

Linearity

$$Z[ax[n] + by[n]] = aX(z) + bY(z), \quad ROC \supseteq (R_x \cap R_y)$$

While it is obvious that the ROC of the linear combination of $x[n]$ and $y[n]$ should be the intersection of the their individual ROCs $R_x \cap R_y$ in which both $X(z)$ and $Y(z)$ exist, note that in some cases the ROC of the linear combination could be larger than $R_x \cap R_y$. For example, for both $x[n] = a^n u[n]$ and $y[n] = a^n u[n-1]$, the ROC is $|z| > |a|$, but the ROC of their difference $a^n u[n] - a^n u[n-1] = \delta[n]$ is the entire z-plane.

Time Shifting

$$Z[x[n - n_0]] = z^{-n_0} X(z), \quad ROC = R_x$$

Proof:

$$Z[x[n - n_0]] = \sum_{n=-\infty}^{\infty} x[n - n_0] z^{-n}$$

Define $m = n - n_0$, we have $n = m + n_0$,

$$\sum_{m=-\infty}^{\infty} x[m] z^{-m} z^{-n_0} = z^{-n_0} X(z)$$

The new ROC is the same as the old one except the possible addition/deletion of the origin or infinity as the shift may change the duration of the signal.

Time Expansion

$$Z[x[n/k]] = X(z^k), \quad ROC = R_x^{1/k}$$

The discrete signal $x[n]$ cannot be continuously scaled in time as n has to be an integer (for a non-integer n $x[n]$ is zero). Therefore $x[n/k]$ is defined as:

$$x[n/k] \triangleq \begin{cases} x[n/k] & \text{if } n \text{ is a multiple of } k \\ 0 & \text{else} \end{cases}$$

Example: If $x[n]$ is ramp.

n	1	2	3	4	5	6
$x[n]$	1	2	3	4	5	6

Then the expanded version $x[n/2]$ is:

n	1	2	3	4	5	6
$n/2$	0.5	1	1.5	2	2.5	3
m		1		2		3
$x[n/2]$	0	1	0	2	0	3

where m is the integer part of n/k.

Proof: The z-transform of such an expanded signal.

$$Z[x[n/k]] = \sum_{n=-\infty}^{\infty} x[n/k]z^{-n} = \sum_{m=-\infty}^{\infty} x[m]z^{-km} = X(z^k)$$

Note that the change of the summation index from n to m has no effect as the terms skipped are all zeros.

Convolution

$$Z[x[n] * y[n]] = X(z)Y(z), \quad ROC \supseteq (R_x \cap R_y)$$

The ROC of the convolution could be larger than the intersection of R_x and R_y, due to the possible pole-zero cancellation caused by the convolution.

Time Difference

$$Z[x[n] - x[n-1]] = (1 - z^{-1})X(z), \quad ROC = R_x$$

Proof:

$$Z[x[n] - x[n-1]] = X(z) - z^{-1}X(z) = (1 - z^{-1})X(z) = \frac{z-1}{z}X(z)$$

Note that due to the additional zero $z=1$ and pole $z=0$, the resulting ROC is the same as R_x except the possible deletion of $z=0$ caused by the added pole and/or addition of $z=1$ caused by the added zero which may cancel an existing pole.

Time Accumulation

$$Z[\sum_{k=-\infty}^{n} x[k]] = \frac{1}{1-z^{-1}} X(z), \quad ROC \supseteq [R_x \cap (|z|>1)]$$

Proof: The accumulation of $x[n]$ can be written as its convolution with $u[n]$:

$$u[n]*x[n] = \sum_{k=-\infty}^{\infty} u[n-k]x[k] = \sum_{k=-\infty}^{n} x[k]$$

Applying the convolution property, we get:

$$Z[\sum_{k=-\infty}^{n} x[k]] = Z[u[n]*x[n]] = \frac{1}{1-z^{-1}} X(z)$$

As $Z[u[n]] = 1/(1-z^{-1})$.

Time Reversal

$$Z[x[-n]] = X(1/z) \quad ROC = 1/R_x$$

Proof:

$$Z[x[-n]] = \sum_{n=-\infty}^{\infty} x[-n]z^{-n} = \sum_{m=-\infty}^{\infty} x[m](\frac{1}{z})^{-m} = X(1/z)$$

where $m=-n$.

Scaling in Z-domain

$$Z[a^n x[n]] = X\left(\frac{z}{a}\right), \quad ROC = |a|R_x$$

Proof:

$$Z[a^n x[n]] = \sum_{n=-\infty}^{\infty} x[n]\left(\frac{z}{a}\right)^{-n} = X\left(\frac{z}{a}\right)$$

In particular, if $a = e^{j\omega_0}$, the above becomes:

$$Z[e^{jn\omega_0} x[n]] = X(e^{-j\omega_0} z) \quad ROC = R_x$$

The multiplication by $e^{-j\omega_0}$ to z corresponds to a rotation by angle ω_0 in the z-plane, i.e., a frequency shift by ω_0. The rotation is either clockwise ($\omega_0 > 0$) or counter clockwise ($\omega_0 > 0$) corresponding to, respectively, either a left-shift or a right shift in frequency domain. The property is essentially the same as the frequency shifting property of discrete Fourier transform.

Conjugation

$$Z[x^*[n]] = X^*(z^*), \qquad ROC = R_x$$

Proof: Complex conjugate of the z-transform of $x[n]$ is:

$$X^*(z) = [\sum_{n=-\infty}^{\infty} x[n]z^{-n}]^* = \sum_{n=-\infty}^{\infty} x^*[n](z^*)^{-n}$$

Replacing z by z^*, we get the desired result.

Differentiation in Z-domain

$$Z[nx[n]] = -z\frac{d}{dz}X(z), \quad ROC = R_x$$

Proof:

$$\frac{d}{dz}X(z) = \sum_{n=-\infty}^{\infty} x[n]\frac{d}{dz}(z^{-n})\sum_{n=-\infty}^{\infty}(-n)x[n]z^{-n-1} = \frac{-1}{z}\sum_{n=-\infty}^{\infty} nx[n]z^{-n}$$

i.e.,

$$Z[nx[n]] = -z\frac{d}{dz}X(z)$$

Example: Taking derivative with respect to z of the right side of:

$$Z[a^n u[n]] = \frac{1}{1-az^{-1}} \qquad |z|>|a|$$

We get,

$$\frac{d}{dz}\left[\frac{1}{1-az^{-1}}\right] = \frac{-az^{-2}}{(1-az^{-1})^2}$$

Due to the property of differentiation in z-domain, we have:

$$Z[na^n u[n]] = \frac{az^{-1}}{(1-az^{-1})^2} \qquad |z|>|a|$$

Note that for a different ROC $|z|<|a|$, we have:

$$Z[-na^n u[-n-1]] = \frac{az^{-1}}{(1-az^{-1})^2} \quad |z|<|a|$$

Pole Zero Plot

In mathematics, signal processing and control theory, a pole–zero plot is a graphical representation of a rational transfer function in the complex plane which helps to convey certain properties of the system such as:

- Stability.
- Causal system/anticausal system.
- Region of convergence (ROC).
- Minimum phase/non minimum phase.

A pole-zero plot shows the location in the complex plane of the poles and zeros of the transfer function of a dynamic system, such as a controller, compensator, sensor, equalizer, filter, or communications channel. By convention, the poles of the system are indicated in the plot by an X while the zeros are indicated by a circle or O

A pole-zero plot can represent either a continuous-time (CT) or a discrete-time (DT) system. For a CT system, the plane in which the poles and zeros appear is the s plane of the Laplace transform. In this context, the parameter *s* represents the complex angular frequency, which is the domain of the CT transfer function. For a DT system, the plane is the z plane, where *z* represents the domain of the Z-transform.

Continuous-time Systems

In general, a rational transfer function for a continuous-time LTI system has the form:

$$H(s) = \frac{B(s)}{A(s)} = \frac{\sum_{m=0}^{M} b_m s^m}{s^N + \sum_{n=0}^{N-1} a_n s^n} = \frac{b_0 + b_1 s + b_2 s^2 + \cdots + b_M s^M}{a_0 + a_1 s + a_2 s^2 + \cdots + a_{(N-1)} s^{(N-1)} + s^N}$$

where,

- B and A are polynomials in s,
- M is the order of the numerator polynomial,
- b_m is the m-th coefficient of the numerator polynomial,

- N is the order of the denominator polynomial,
- a_n is the n-th coefficient of the denominator polynomial.

Either M or N or both may be zero, but in real systems, it should be the case that ; otherwise the gain would be unbounded at high frequencies.

Poles and Zeros

- The zeros of the system are roots of the numerator polynomial:

$$s = \{\beta_m \mid m \in 1,\ldots M\} \text{ such that } B(s)|_{s=\beta_m} = 0$$

- The poles of the system are roots of the denominator polynomial:

$$s = \{\alpha_n \mid n \in 1,\ldots N\} \text{ such that } A(s)|_{s=\alpha_n} = 0$$

Region of Convergence

The region of convergence (ROC) for a given CT transfer function is a half-plane or vertical strip, either of which contains no poles. In general, the ROC is not unique, and the particular ROC in any given case depends on whether the system is causal or anti-causal.

- If the ROC includes the imaginary axis, then the system is bounded-input, bounded-output (BIBO) stable.
- If the ROC extends rightward from the pole with the largest real-part (but not at infinity), then the system is causal.
- If the ROC extends leftward from the pole with the smallest real-part (but not at negative infinity), then the system is anti-causal.

The ROC is usually chosen to include the imaginary axis since it is important for most practical systems to have BIBO stability.

Example:

$$H(s) = \frac{25}{s^2 + 6s + 25}$$

This system has no (finite) zeros and two poles:

$$s = \alpha_1 = -3 + 4j$$

and

$$s = \alpha_2 = -3 - 4j$$

The pole-zero plot would be:

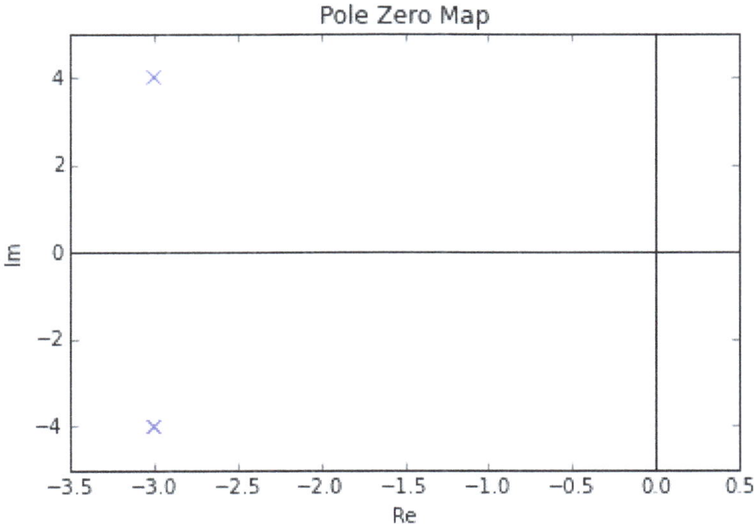

Notice that these two poles are complex conjugates, which is the necessary and sufficient condition to have real-valued coefficients in the differential equation representing the system.

Discrete-time Systems

In general, a rational transfer function for a discrete-time LTI system has the form:

$$H(z) = \frac{P(z)}{Q(z)} = \frac{\sum_{m\ 0} b\ z}{\sum a\ z} = \frac{b_0 + b_1 z^{-1} + b_2 z^{-2} \quad + b_M z^{-}}{1 - a_1 z^{-1} + a_2 z^{-2} \quad + a\ \ z^{-}}$$

where,

- M is the order of the numerator polynomial,
- b_m is the m-th coefficient of the numerator polynomial,
- N is the order of the denominator polynomial,
- a_n is the n-th coefficient of the denominator polynomial.

Either M or N or both may be zero.

Poles and Zeros

- $z = \beta_m$ such that $P(z)|_{z=\beta_m} = 0$ are the zeros of the system.
- $z = \alpha_n$ such that $Q(z)|_{z=\alpha_n} = 0$ are the poles of the system.Region of convergence.

The region of convergence (ROC) for a given DT transfer function is a disk or annulus which contains no poles. In general, the ROC is not unique, and the particular ROC in any given case depends on whether the system is causal or anti-causal.

- If the ROC includes the unit circle, then the system is bounded-input, bounded-output (BIBO) stable.
- If the ROC extends outward from the pole with the largest (but not infinite) magnitude, then the system has a right-sided impulse response. If the ROC extends outward from the pole with the largest magnitude and there is no pole at infinity, then the system is causal.
- If the ROC extends inward from the pole with the smallest (nonzero) magnitude, then the system is anti-causal.

The ROC is usually chosen to include the unit circle since it is important for most practical systems to have BIBO stability.

Example:

If $P(z)$ and $Q(z)$ are completely factored, their solution can be easily plotted in the z-plane. For example, given the following transfer function:

$$H(z) = \frac{z+2}{z^2 + \frac{1}{4}}$$

The only (finite) zero is located at: $z = -2$, and the two poles are located at: $z = \pm\frac{j}{2}$, where j is the imaginary unit.

The pole–zero plot would be:

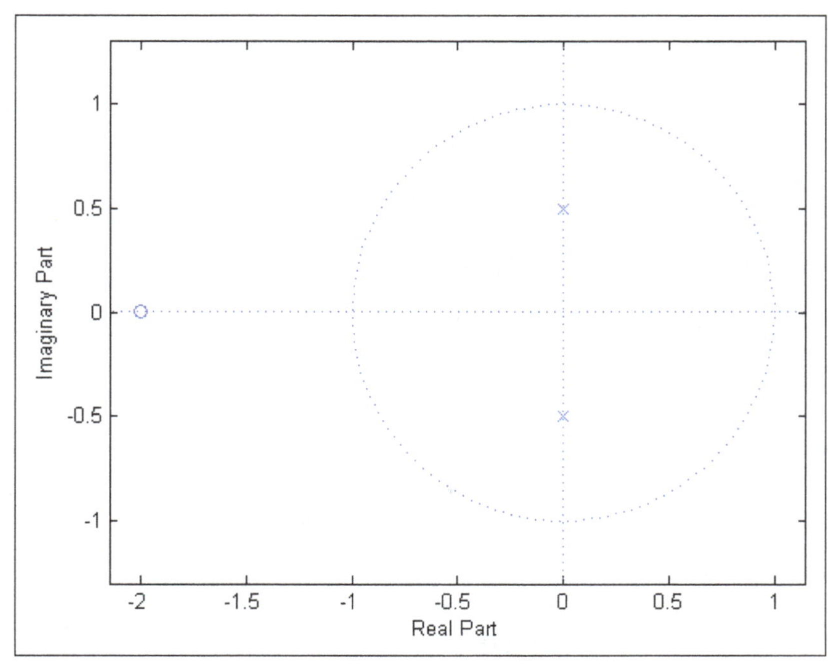

Inverse Z-transfrom

If we want to analyze a system, which is already represented in frequency domain, as discrete time signal then we go for Inverse Z-transformation.

Mathematically, it can be represented as,

$$x(n) = Z^{-1}X(Z)$$

where xn is the signal in time domain and X z is the signal in frequency domain.

If we want to represent the above equation in integral format then we can write it as

$$x(n) = (\frac{1}{2\Pi j})\oint X(Z)Z^{-1}dz$$

Here, the integral is over a closed path C. This path is within the ROC of the x z and it does contain the origin.

Methods to Find Inverse Z-Transform

When the analysis is needed in discrete format, we convert the frequency domain signal back into discrete format through inverse Z-transformation. We follow the following four ways to determine the inverse Z-transformation.

- Long Division method.
- Partial Fraction expansion method.
- Residue or Contour integral method.

Long Division Method

In this method, the Z-transform of the signal x z can be represented as the ratio of polynomial as shown below:

$$x(z) = N(Z)/D(Z)$$

Now, if we go on dividing the numerator by denominator, then we will get a series as shown below:

$$X(z) = x(0) + x(1)Z^{-1} + x(2)Z^{-2} + \dots \quad \dots \quad \dots$$

The above sequence represents the series of inverse Z-transform of the given signal $forn \geq 0$ and the above system is causal.

However for n<o the series can be written as:

$$x(z) = x(-1)Z^{1} + x(-2)Z^{2} + x(-3)Z^{3} + \dots \quad \dots \quad \dots$$

Partial Fraction Expansion Method

Here also the signal is expressed first in N z /D z form.

If it is a rational fraction it will be represented as follows:

$$x(z) = b_0 + b_1 Z^{-1} + b_2 Z^{-2} + \ldots \quad \ldots \quad \ldots + b_m Z^{-m})$$
$$/(a_0 + a_1 Z^{-1} + a_2 Z^{-2} + \ldots \quad \ldots \quad \ldots + a_n Z^{-N})$$

The above one is improper when m < n and an ≠ o.

If the ratio is not proper i.e., improper, then we have to convert it to the proper form to solve it.

Residue or Contour Integral Method

In this method, we obtain inverse Z-transform xn by summing residues of $[x(z)Z^{n-1}]$at all poles. Mathematically, this may be expressed as:

$$x(n) = \sum_{all \quad poles \quad X(z)} {}^{residues \quad of[x(z)Z^{n-1}]}$$

Here, the residue for any pole of order m at $z = \beta$ is:

$$Residues = \frac{1}{(m-1)!} \lim_{Z \to \beta} \{\frac{d^{m-1}}{dZ^{m-1}} \{(z-\beta)^m X(z) Z^{n-1}\}.$$

Permissions

All chapters in this book are published with permission under the Creative Commons Attribution Share Alike License or equivalent. Every chapter published in this book has been scrutinized by our experts. Their significance has been extensively debated. The topics covered herein carry significant information for a comprehensive understanding. They may even be implemented as practical applications or may be referred to as a beginning point for further studies.

We would like to thank the editorial team for lending their expertise to make the book truly unique. They have played a crucial role in the development of this book. Without their invaluable contributions this book wouldn't have been possible. They have made vital efforts to compile up to date information on the varied aspects of this subject to make this book a valuable addition to the collection of many professionals and students.

This book was conceptualized with the vision of imparting up-to-date and integrated information in this field. To ensure the same, a matchless editorial board was set up. Every individual on the board went through rigorous rounds of assessment to prove their worth. After which they invested a large part of their time researching and compiling the most relevant data for our readers.

The editorial board has been involved in producing this book since its inception. They have spent rigorous hours researching and exploring the diverse topics which have resulted in the successful publishing of this book. They have passed on their knowledge of decades through this book. To expedite this challenging task, the publisher supported the team at every step. A small team of assistant editors was also appointed to further simplify the editing procedure and attain best results for the readers.

Apart from the editorial board, the designing team has also invested a significant amount of their time in understanding the subject and creating the most relevant covers. They scrutinized every image to scout for the most suitable representation of the subject and create an appropriate cover for the book.

The publishing team has been an ardent support to the editorial, designing and production team. Their endless efforts to recruit the best for this project, has resulted in the accomplishment of this book. They are a veteran in the field of academics and their pool of knowledge is as vast as their experience in printing. Their expertise and guidance has proved useful at every step. Their uncompromising quality standards have made this book an exceptional effort. Their encouragement from time to time has been an inspiration for everyone.

The publisher and the editorial board hope that this book will prove to be a valuable piece of knowledge for students, practitioners and scholars across the globe.

Index

www.ingramcontent.com/pod-product-compliance
Lightning Source LLC
Chambersburg PA
CBHW082057190326
41458CB00010B/3516

* 9 7 8 1 6 4 1 7 2 3 9 8 5 *